Contents

Tables

Figures

Introduction

For thousands of years, the principal activity of man was the procurement of food. More recently, with the agrarian revolution, which preceded the industrial revolution of the last century, social problems associated with the creation of large industrial enterprises and urbanization seemingly became the principal concern. The terrible inflation in agricultural prices which occurred between 1972 and 1975 once again raised the possibility of a major food crisis in this century. Will the concerns expressed by economists such as Malthus, comparing the growth rate of agricultural production with that of population, prove justified?

If this question is worrisome for solvent, food-importing countries such as Japan, Europe, and the OPEC countries, it is agonizing for developing countries that lack petroleum resources because for them it can be reduced to a single tragic word: famine.

Since June of 1973, when the United States startled the world by decreeing an export embargo on soybeans, its increasing power to influence human nutrition in the next century has become an additional concern. Must the European Economic Community (EEC) and Japan, already subject to pressure from the oil-producing countries, now contend with the agricultural power of grain and protein crop producers?

A response to both of these concerns is of vital significance. Our conclusions in this book are very clear: the similarities between OPEC's power in energy and the power of the United States to control food production are real. The performance and methods of American agriculture, therefore, merit careful analysis.

We have based our conclusions on data from the most serious studies available and have carefully examined the hypotheses on which they are based. This task has not been easy because interest in these questions on the world level is so recent: research on this subject began only in the 1970s.

Our investigation has therefore concentrated on the heart of the food chain: the grain-oilseed-livestock (GOL) complex. These products compose the bulk of world commerce and are the basis for all secondary products:

flour, livestock feed, beef, pork, poultry, and dairy products. Granted, this approach does not permit the study of such important products as fruits, vegetables, wines, and alcoholic beverages, nor of tropical products such as coffee, cocoa, and tea, nor of certain Mediterranean products such as mutton or olive oil. It was necessary, however, to limit our study to products of strategic importance in world human and animal nutrition.

The works we have studied on the grain-oilseed-livestock complex generally provided projections through the year 1985. It appears that their principal conclusions, however, can be extended through the year 1990 and even through the rest of the century. Indeed, it is now agreed that world population growth has stabilized at less than 2 percent per year. Furthermore, the productivity reserves of farmers in both industrialized and developing countries are greater than initially indicated by analyses begun in 1974. These two inflections allow us to confirm, for 1990 and 2000, the conclusions drawn beginning in 1974 for the year 1985: global balance will be achieved, with certain constraints for some developing countries, particularly in Africa, and for meat consumption. This balance, of course, assumes the absence of a major world incident halting North-South and East-West trade flows. Peace is essential for the difficult task of adjusting agricultural supplies to world demand.

Readers wishing to delve into this major problem and its effects on the future of mankind will find in Part I a discussion of the evolution of agricultural equilibrium at the world level (Chapter 1), at the European level (Chapter 2), and at the U.S. level (Chapter 3). The central role played by the United States will thus be demonstrated, notwithstanding the importance of its Brazilian and Argentinian "challengers" or that of the old colonial powers, owing to the inescapable needs of the Eastern bloc countries.

Part II will deal with the agricultural situation and policies in the United States (Chapter 4) because the efficiency and reaction capacity of this complex (Chapter 5) merits fresh examination, free of currently held preconceptions. At a time when the energy crisis seems to have profoundly affected the resilience of the American economy, its agricultural sector has taken on a key role in the nation's future (Chapter 6).

The future of American agriculture is also the future of Europe, for two reasons: directly, because Europe has not said its last word in the area of agriculture, and indirectly, because concern for adequate nutrition of the developing countries must be one of the principal objectives of the international community for the rest of this century.

Abbreviations

CAP   common agricultural policy of the EEC

$ U.S. dollar

EEC   European Economic Community

ERS   Economic Research Service of the USDA

GATT General Agreement on Tariff and Trade

GOL   Grain-oilseed-livestock model of the USDA

HRW Hard red winter wheat

INRA Institut National de la Recherche Agronomique

Mha   Millions of hectares (1 ha = 2.5 acres)

Mt   Millions of metric tons (1 metric ton = 2,217 pounds = 1,000 kilograms)

NAS   National Academy of Sciences

OPEC Organization of Petroleum Exporting Countries

PIK   Payment in kind

PL 480 Agricultural Trade Development and Assistance Act of 1954 (Public Law 480)

USDA U.S. Department of Agriculture

Part One

THE WORLD FOOD AND AGRICULTURAL BALANCE

1

Global Balance and Partial Imbalances through the Year 2000

Two political events during the course of 1972 were at the root of what some have called the great food crisis. This crisis has in turn drawn world attention to the role of agriculture, a subject that previously had not generated a level of interest commensurate with the problem.

The first event was the Soviet-American rapprochement in 1972, through which the Soviet Union agreed to become a purchaser of grain on the American market. This was a position the Soviets had carefully avoided in the past, relying instead on a limited number of contracts with Canada.

The second event occurred the same year: a new secretary of agriculture was appointed by President Richard Nixon (the previous secretary, Clifford Harding, had been blamed for the Republican party's loss of farm votes in the Midwest). Earl Butz, the new secretary, carried out a policy of financial incentives to American farmers to reduce the amount of land under cultivation, thereby decreasing stocks and increasing prices on the grain market. Five million hectares were thus withdrawn from cultivation, corresponding to a voluntary decrease in production of 20 million metric tons of grain.

Two apparently unrelated events had considerable repercussions in the agricultural markets. First, a bad harvest led the Soviets to make massive purchases from the United States. American grain exports to all countries increased from 44 million metric tons in 1971–72 to 73 million metric tons in 1972–73, of which 14 million metric tons were exported to the Soviet Union.

Compounding this momentary imbalance was the 1974 drought in the Corn Belt, which reduced corn yields by one-fourth. The price of corn, which had already increased by 45 percent between 1971 and 1972, climbed by another 62 percent from 1972 to 1973 and by 19 percent from 1973 to 1974. All told, the prices for American grains almost tripled in three years. In retrospect, it can be seen that the advent of "green gold" came a full year before the creation of "black gold" by the oil-producing countries.

In 1973, for the first time in recent memory, panic swept through the ranks of world agricultural experts. Was this the onset of a new period of famine? The world's vulnerability to climatic aberrations appeared much greater than previously imagined. Projections would have to be recalculated and the future reexamined in the light of recent developments in world grain commerce.

The World Food Conference, which convened November 5–16, 1974, came as a result of these events. Proposed by Secretary of State Henry Kissinger, this initiative—not without ulterior motives—corresponded to a growing awareness of agricultural problems within the political establishment.

The politicians soon discovered that no detailed projections had been conducted on world food prospects for the remainder of the twentieth century. This lack of any serious attempt at econometric modeling was particularly significant in the area of agriculture. The question raised by these politicians was indeed of great importance: Was the situation in 1973 a mere accident or the portent of a profound crisis in food production during the last quarter of the century? Two opposing methods of reasoning were brought to bear on this question.

The first method, in fashion for a time, was based on the experts' intuition. It consisted of projecting, based on current trends, the principal factors affecting the supply of and demand for agricultural products: population, income, caloric or protein needs, usable land surface, and other factors. It was concluded that in the medium term, food production would be inadequate to assure a decent per capita caloric ration to the world's population. This situation is analogous to the energy situation. With'n this frame of reference the realities of the agricultural markets (the deteı.nination of relative prices for different products) lose their importance. The prime consideration is to find substitutes for existing products and producers, regardless of the costs.

A second method of reasoning began to emerge, which arrived at the opposite conclusion: certain crops will be produced in excess. This overproduction will result in price modifications (either in the marketplace or through government intervention), which in turn will influence demand. World supply and demand will therefore stabilize at a determined price level. Even a slight variation in demand will cause a change in price, which in turn will modify supply and so on.[1]

It is useful to examine the reasons why this method of analysis, which recognizes the interdependence among economic aggregates, took so long to emerge. To analyze this situation correctly, it is indispensable to have an econometric model capable of explaining the behavior of the world's principal producers, specifically those that can modify their production in response to world demand. The construction of such models depicting a structure of production is difficult enough for a single product in the short term. It becomes more complex when the study covers the long term because the effect of technological advances must be foreseen. It is also necessary to consider a wide range of complementary crops because a given farming operation can produce a variety of vegetable staples or can adjust to accommodate certain forms of livestock production to a greater or lesser extent according to market conditions.

Finally, beyond the technical difficulties revealed by the construction of national macroeconomic models, there remained a final obstacle to overcome. When the econometrist constructs a model, he must make a number of choices. The most crucial among these concerns the purpose of his efforts: should he construct a model offering the best possible price forecasts, or, inversely, the best possible explanation of structural adjustments (geographic distribution of crops, specialization, and so on) at a given price level?

In its current state of development, the science of econometrics does not permit these two approaches to be combined in a single model, even an extensive one. It is understandable, under these conditions, why the governments and the various institutions concerned with the future of agriculture hesitated until 1973 to finance these studies.

RESULTS OF AGRICULTURAL SUPPLY
AND DEMAND PROJECTIONS

For some time economists had wanted to undertake a comprehensive agricultural econometric study, but until 1973 they had been prevented from doing so by the high cost of such models. The food crisis of 1972–74 enabled them to realize their ambitions. We have been able to locate six of these studies, the results of which are shown in Table 1.1.

TABLE 1.1. Forecast of the difference between supply and demand in 1985 (in millions of tons of grains or meat)

	World		Developing Countries		Industrialized Countries	
	Grains	Meat	Grains	Meat	Grains	Meat
USDA[a]						
I	0	+0.60	−48.9	+0.60	+48.9	+0.01
II	0	+0.60	−70.6	+0.70	+70.6	−0.10
III	+3.2	+0.60	−41.0	+0.10	+44.2	+0.50
IV	+3.1	+0.50	−34.1	+0.30	+37.2	+0.20
California	−0.6	−1.20	−90.9	+0.60	+90.3	−1.80
OECD[b]	+38	−1.50	−48.5	+0.60	+86.5	−2.10
	+56		−37.5		+93.5	
FAO			−85			
Iowa[b]	+144		−66			
	+108		−113			
IFPRI[b]			−66			
			−83			

Note: A plus sign (+) indicates an exportable surplus for the zone; a minus sign (−) implies a deficit that has to be compensated by imports, food aid, and restrictions.

[a]I, II, III, and IV fit the four scenarios used by USDA researchers to work their GOL model. Schematically, they correspond to the following: I—status quo hypothesis; II—free trade hypothesis; III—protectionist hypothesis; IV—green revolution hypothesis.

[b]The OECD, Iowa, and IFPRI studies provide two extreme figures for grains.

Two of these studies are limited to quantitative summaries: a study by FAO in 1974 and another by the University of California.[2] Conceptually, they follow rather closely the "trend projection" method, although FAO subsequently developed an econometric model. Two others are more economic in nature: a study by OECD and another by the U.S. Department of Agriculture.[3] Only the results of the last two studies, a report by the University of Iowa and another by the International Food Policy Research Institute, are obtainable.[4] We have been unable to obtain the complete studies.

The consequences of these different approaches can be seen by examining the overall results of the studies cited, which are shown in Table 1.1. One of the studies forecasts a food deficit in the developing countries (FAO), without addressing the world food imbalance. Two others (USDA and the University of California) forecast that in 1985, supply will precisely equal demand. The sum of OECD's forecasts for each region shows a large worldwide surplus.

The first conclusion that can be drawn is that unanimity is difficult to achieve. The first alarmist movement was launched by FAO, which forecast a deficit of 85 million metric tons of grain for the developing countries. This international organization was aiming primarily to attract attention to the size of a presumed food deficit and the difficulties in remedying it without price increases too great for the national budgets of the poorest countries. But it is also true that the supply projections made by the FAO were very imprecise and that the impression of widespread malnutrition resulted more from intuition than from a serious comparison of supply and demand.[5] Since that time, FAO has conducted complementary studies on supply and seems to concur with the more commonly held opinions, which can be summarized as follows:

1. *There will not be a global food deficit in 2000. On the contrary, there could even be a large surplus.* This applies more to grain than to meat, an important point that will be covered in more detail further on.

2. *The deficit is going to worsen for developing countries.* Yet even within this group of nations, significant disparities will emerge, based on solvency (the OPEC countries) or the degree of success of agricultural policies (Brazil, Argentina, Thailand, and India, for example).

3. *The surplus in the industrialized countries will increase,* but here again the trend will not be consistent for all nations in this group. The economic consequences of these disparities are at least as significant as those for the preceding group.

Studies completed during the 1980s confirmed these three conclusions. Noteworthy in this regard are the ten-year projection model (1993) by Winrock International and *Food and Fiber Projections to 2000* by Resources for the Future, which are based on trend projections calculated by a group of experts and include coherence functions related to production and consumption. Comparison of these projections allows an evaluation of trade among twelve regions of the world.[6] The same is true of the Vienna model, prepared at the request of OECD by IASA in cooperation with Michigan State University. The results of this study will be available in 1986. The International Wheat Council's *Long-Term Grain Outlook,*[7] initiated in 1979, forecasts production and trade in 2000 for wheat, rice, and coarse grain. The world is divided into six regions: low-income developing countries, other developing countries, USSR and Eastern Europe, China, six major exporting countries, and other developed countries. The deficit in total grains of low-income countries is forecast at 61 million metric tons (Mt) and that of other developing countries at 76 Mt, making a total 137 Mt deficit for all developing countries.

With respect to the global situation, an initial observation should be made about the problem of simultaneity. Certain studies forecast a surplus of approximately 50 Mt (OECD) or even 100 Mt (Iowa) of grain. Others foresee an equilibrium in which supply equals demand, a result that stems from the construction of an equilibrium model. Here the approach that favors price as the regulating factor, as opposed to the projection method, offers significant advantages. Indeed, in both cases grain production is channeled through economic circuits, with prices playing an essential role because they will determine the market for meat, an extremely important factor in the world grain market.

To illustrate the role of prices, it is useful to consider a fundamental development which dramatically changed the nature of world food problems: increased consumption of meat. Until recent times, grains, primarily wheat, rice, and rye, were cultivated essentially for human consumption. With the extraordinary development of corn cultivation and the introduction of soybeans as a complementary source of protein, livestock feed progressively surpassed human consumption as the world's principal use of grain. In the industrialized countries, a large portion of grain production and imports is used, in conjunction with soybeans or other vegetable proteins, for the production of pork, poultry, and, to a lesser degree, beef. In addition to corn, barley, oats, and soybeans, a part of the world's wheat and rice harvests (bran and portions of low quality) is also used for animal feed. In the developing countries, rapid urbanization and changes in the dietary habits of the wealthier classes have resulted in large imports of meat. Local authorities are attempting to replace these imports with locally produced livestock raised on imported grain and soybeans.

Under these circumstances it becomes clear why prices will play a major role in determining the allotment of grain between human and livestock consumption.

This relationship between the grain and meat markets is the subject of in-depth analysis in the USDA's GOL model. Its purpose is to project the effect on prices of different international commercial strategies and, in turn, the effect of prices on quantities produced in the various regions of the world.

The GOL model bases its projections on four principal scenarios:

Scenario I (status quo) postulates a continuation of the principal national strategies for each large group of countries. Protectionist policies and trade barriers currently in effect are maintained, but complete self-sufficiency is not attained. A certain permeability in trade continues.

Scenario II (free trade) is based on high world growth rates with strong growth in demand for imports, which is facilitated by the elimination of current protectionist practices.

Scenario III (protectionism) hypothesizes a generalized deceleration in growth. Imports decrease significantly and protectionist policies are widely enforced in a substantially more aggressive manner than in Scenario I.

Scenario IV (green revolution) tests the hypothesis of a vigorous acceleration in the productivity of the developing countries, where the green revolution is assumed to have conclusively succeeded.

These four scenarios are summarized in Table 1.1 for grains and meat.

The GOL model also shows a substantial increase in the price ratio of beef to corn and a much more limited increase in the price ratio of pork to corn for all the industrialized zones, with the exception of the EEC, where the price policy is moderating this trend.

A number of serious problems result from this price ratio increase because decisions concerning grain production decisively affect meat producers. Indeed, the livestock sector is handicapped in two ways which grain producers are not. First, the production cycles are much longer for meat. It takes at least three years to regenerate a herd of cattle and at least eighteen months for a herd of hogs. Second, meat producers are much more vulnerable to fluctuations in demand, and consumer demand is more sensitive to price variations in meat than in grain. Countries that export meat or livestock feed will thus be competing in a more uncertain and interdependent world.

The Kellogg Foundation, with Kenneth Farrell, formerly of the USDA, contributed to the revision of GOL and other models used by the USDA since 1973.[8] The "remodeling" of GOL is under way and resulted in publication of *World Food Study for the Year 2000*, using 1982 as the new base year.[9] This revised model introduces an annual tracking of the value of the dollar, the standard against which the various world prices are measured. It adds two new scenarios: widespread decline in production following climatic shocks and increase in the revenues of the developing countries.

Based on the GOL model scenarios and our previous reflections on the other studies, we can refine our three earlier, very general conclusions:

1. It is practically certain that there will be no global grain deficit during the next ten years or even afterward.

2. Based on the current price relationship between meat and corn, there will probably be a meat deficit. An increase of 30 percent in this ratio in the large meat-producing countries (the United States, Canada, and Oceania) would appear to assure equilibrium in supply and demand.

3. It is improbable that the developing countries as a whole, and particularly Africa, will attain self-sufficiency in grain. If they cannot generate enough exports to enable them to finance their needs in grain, given the current market structure, the problem of food aid will become increasingly acute in the years ahead.

4. If the growth rates in the revenues of the importing countries (developed or not) are not high enough, these countries will be tempted to isolate them-

selves behind customs barriers. This would provide a means to minimize their deficits, but the social cost would be very high for the exporting countries, which would face a choice between supporting grain prices or decreasing production.

5. Global agreements covering wheat, feed grains, and meat are indispensable. To create a rough outline for such an agreement, it is necessary to evaluate rather precisely the long-term needs for meat in the developing countries. To our knowledge, such an evaluation has not been conducted, although there are several reports on the industrialized countries. The economic strategies for grain and meat are highly interdependent, with price as the mediating factor.

CONDITIONS NECESSARY FOR WORLD FOOD EQUILIBRIUM

The relatively optimistic conclusions of the different scenarios posed by the American study are conditioned on a certain number of fundamental hypotheses.

Any supply/demand projection for agricultural products must involve four essential factors: on the demand side, change in population and income; on the supply side, changes in the area of cropland under cultivation and in the yield per acre. Most of the studies and econometric models treat these changes as exogenous, that is, determined outside of the model, which strictly speaking is a questionable approach. After all, it cannot be ignored that the population of a country depends on the amount of food available, and the amount of land under active cultivation depends on the yields obtained, the relative price of the product, and the cost of production. This must not, however, prevent us from raising the primary question: Are these relatively optimistic projections of world food supply and demand conditioned on the realization of the hypotheses concerning the exogenous factors?

The annual rate of growth in demand for food is determined by *the rate of population growth*. Estimates of world population have been revised downward by the statisticians of the United Nations,[10] and the rate of world population growth is decreasing steadily—from 1.94 percent per year in 1970 to 1.72 percent in 1980, 1.65 percent in 1990, 1.50 percent in 2000, and 1.07 percent in 2020. This rate, they contend, will approach zero by 2100, but in the year 2000, it will still represent an annual increase of 90 million inhabitants in the world's population. Based on this growth rate, world population would reach slightly less than 5 billion in 1985, a little more than 6 billion in 2000, and would stabilize at about 10.5 billion by the end of the next century. By comparison, the evaluations generally accepted a few years ago were based on a growth rate of 2 percent or slightly more per year. The average figure of 1.72 percent in 1980 encompasses a wide disparity between the growth rate in the industrialized countries (including Eastern Europe), estimated at 0.5 percent per year in 1990 (down from 1.1 percent in 1970), and the growth rate in the developing countries, estimated at 1.9 percent (down from 2.5 percent in 1970). Based on these rates, the population of the industrialized countries will increase from 1.1 billion inhabitants in 1982 to 1.2 billion in 2000, while the population of the developing countries will increase from 3.4 billion in 1982 to 4.0 billion in 1990 and 4.8 billion in 2000.

Experience has shown in recent years that as long as the annual population growth rate in the developing countries does not surpass 2.9 percent, agricultural production should stay ahead of population growth, despite the limited resources deployed in those countries (see Appendix 1). The nutritional needs in energy and protein would thus, theoretically, be satisfied in the developing countries, with the exception of a number of African countries.

At this point, it is useful to mention a few words on a subject that will be addressed in greater detail in our discussion of economic hypotheses: the question of *income distribution*.

The optimism arising from the relative slowdown in population growth does not, unfortunately, dispel a profound pessimism over the fate of a large portion of the world's population. As shown in a report presented by the president of the World Bank at the Belgrade Conference in October of 1979, approximately 800 million people throughout the world live in a state of absolute poverty.[11] Of this total, approximately 500 million suffer from nutritional deficiencies resulting from their chronic state of undernourishment.

Under such conditions is it possible, with any decency, to speak of world nutritional balance, of satisfied needs? Obviously not. The remarks that have been made on this subject must be interpreted differently. They demonstrate simply that a decrease in the rate of population growth permits, perhaps for the first time, a balance in the world's supply of and demand for food. The task of channeling food from areas of abundance to the populations who need it most must still be accomplished. It is not contradictory to affirm that since 1977 India has attained self-sufficiency in grain while at the same time part of its population is suffering from malnutrition. It is clear that only through deliberate, determined efforts will governments be able to distribute grain supplies so as to eliminate hunger.

Unfortunately, as will be shown in our detailed examination of these hypotheses, success will not come easily for the agricultural policies of the developing countries. Maintaining a growth rate of 3 to 4 percent in agricultural production requires a heroic effort and a determination rarely found in the governments concerned.

The world situation in 2000 will thus be characterized by a coexistence of two contradictory phenomena. In the aggregate, world supplies of food will meet world demand, but geographic and social distribution will be difficult to improve. Particularly in the developing countries, a truly adequate food policy will depend above all on the distribution of income among the various socioeconomic classes.

Turning to supply, are the projections of increased production as optimistic as those concerning population growth? All the studies examined here (with the exception of the USDA study) treat three critical hypotheses as entirely exogenous: arable land resources, productivity, and climate. It is simple to demonstrate that, given the current limitations in meteorological control, only the last factor, climate, is truly independent of the economic environment as determined by prices. First, it is necessary to clarify these concepts. Arable land resources constitute a physical factor. Given the technological possibilities for irrigation and water control, it is possible to determine the amount of land that could become available if needed. Calculating production on this basis is more complex, however; it will depend on the intensity with which the necessary inputs are applied. This intensity will, in turn, depend on the unit cost of the inputs in relation to the unit price of outputs. Thus the availability of arable land is only an extremely rough indicator of the potential supply. Nonetheless, it is still useful to attempt to trace the technological boundaries for production. In this respect, the question is the following: In the case of greatest need (that is, of extremely high prices for agricultural products), could the necessary basic resources (inputs) be obtained to assure adequate production? Following is a brief summary of the response provided by the studies of the food situation in 1985.

The amount of *available arable land* is very large. FAO and the University of California go so far as to estimate that only 46 percent of the world's available arable land is currently being cultivated. For a number of countries, however, the maximum limit has been reached. Such is the case, for example, in India, Bangladesh, and Java. Furthermore, according to the OECD, the possibilities for extending cropland of sufficient quality to be suitable for grain production are very limited in some industrialized areas such as Europe and Japan.

The California study indicates that if all available land were cultivated with current average yields, 8 billion people could be fed without difficulty. The assumption of "current average yields" merits greater attention. It is obvious that if yields increase rapidly enough, it would not even be necessary to increase the amount of land under cultivation to feed 6 billion people in 2000. The "increase in yields," however, must be defined with precision. The invariable input, in this case the amount of land under cultivation, determines the scale. Thus, when various figures are factored into the equation as "yield increases," they correspond to a given area of land—the area existing today. That the marginal productivity of supplementary land is generally inferior to that of areas already under cultivation in no way rules out a substantial increase in yields. Any improvement in efficiency, and thus in yields, corresponding to a given fixed input level, will be even greater when the fixed input increases as well, so long as the marginal productivity of the fixed input is positive.[12]

In countries with large reserves of arable land, estimates of yield increases are, by definition, underestimated because when additional land is available for cultivation, the potential for increases in average yield is high. Furthermore, all of the studies agree that an aggregate increase in yield is feasible, whereas an increase in land area under cultivation would presuppose an environment of extremely high agricultural prices.

The California model indicates that if climatic conditions remain constant, an average annual increase of 2 percent in land resources and 2 percent in yields would amply suffice to meet food requirements in the developing countries. This is not currently the case, as shown in the GOL model: land area is increasing by only 1 percent per year and yields by only 1.6 percent. The four scenarios postulated by USDA do not envision land area increases greatly surpassing 1 percent per year in the developing countries. OECD goes so far as to hypothesize that no increase in land area under cultivation will be necessary because productivity gains are potentially of greater significance and easier to put into effect than the development of new cropland. According to the USDA study, area development would occur at a rate of 2 percent per year in the exporting countries only under the free trade scenario (II), or to a lesser extent (1.7 percent per year and exclusively in the developing exporting countries) under the green revolution scenario (IV). The cropland of developed exporting countries such as the United States, Canada, and Australia would even decrease by 0.7 percent per year under the protectionist scenario (III).

Thus increases in supplies will result primarily from technology transfers to the developing countries. A substantial improvement in the techniques of seed selection and conservation after harvest will, in themselves, constitute a great leap forward.[13] Nonetheless, in the world's poorest countries these improvements are slow in getting under way. Recent history has shown these countries lagging far behind and experiencing great difficulty in closing the gap, which, nonetheless, has begun to narrow slowly since 1972. This is shown in Table 1.2, taken from FAO production year books.

The hypotheses postulating potentially *high rates of growth in crop yields* in the developing countries (1.6 percent in the 1970s and 1.9 to 2.9 percent in the 1980s according to the GOL study) are highly aggregated, and although they are consistent on a world scale, an examination of a few particular cases demonstrates their fragility. Certain developing countries, such as India, Pakistan, and Indonesia, are improving their yields, but others are stagnating. This stagnation is dramatic in Africa. We do not possess reliable figures on the yields for this part of the world, but experts with the FAO are currently very pessimistic. This sentiment is based on the following data. In "normal" or "favorable" years, from the standpoint of climate (including, of course, exceptional years such as 1976), it is projected that production will outpace population growth in the developing countries. This highly aggregated method of projection has proven to be more or less accurate, depending on the year in question, for most of the Third World countries. But such is not the case for Africa, as shown in Appendix 1. Africa's lagging performance tends to lower the aggregate statistics for all Third World countries.

This situation should not, however, be cause for excessive alarm. In Table 1.2, taken from the FAO, we can see that an increase of 2 percent per year in the productivity of developing countries would result by 2000 in yields of approximately 22 quintal per hectare (q/ha) for wheat, 22 for corn, and 32 for rice, levels which are greatly inferior to those found in 1980 in the industrialized

TABLE 1.2. Crop yield variations over twenty years

	Wheat			Rice			Corn		
	Variation[a] (%)		Yields,[b]	Variation[a] (%)		Yields,[b]	Variation[a] (%)		Yields,[b]
	1961–63 & 1971–73	1971–73 & 1979–81	1979–81	1961–63 & 1971–73	1971–73 & 1979–81	1979–81	1961–63 & 1971–73	1971–73 & 1979–81	1979–81
Developed Countries									
Market Economy	2.8	0.8	23.6	1.1	−0.1	54.7	3.5	1.8	56.2
USA	2.9	0.5	22.9	2.1	0.2	51.9	4.9	1.5	65.0
France	4.7	1.6	48.9	–	–	–	5.9	0.7	54.5
Italy	2.6	0.9	26.8	–	–	–	–	–	–
Australia	−1.0	1.5	12.6	0.7	−1.0	62.4	–	–	–
Developing Countries									
Market Economy	2.4	2.4	14.7	1.4	1.3	21.5	1.3	1.9	14.9
India	4.5	2.1	15.5	1.4	1.0	18.9	0.6	0.8	11.1
Argentina	0.3	0.3	15.5	–	–	–	2.8	3.9	31.8
Centrally Planned Economy Countries	4.9	1.8	18.3	1.4	3.0	39.0	2.8	1.4	32.0
China	3.9	6.6	19.0	1.4	3.8	42.3	1.6	0.9	30.1
Romania	5.3	2.0	26.1	–	–	–	4.8	3.6	35.7
USSR	5.1	0.0	15.5	–	–	–	3.0	0.3	28.6

Source: FAO, *Production Year Book*, various years.
Note: The average numbers apply to all the countries in each of the three zones. Those cited individually are mentioned only as examples.
[a]Annual growth rate of yields between years given.
[b]Yields are given in quintal per hectare (q/ha); for wheat 1 q/ha = 1.46 bushel per acre; for rice 1.96; for corn 1.58.

13

countries (24, 56, and 55 q/ha respectively). There is therefore considerable potential for progress on a purely technological level which is not hypothetical.

Is the potential for growth in crop yields as high in the industrialized countries? In a recent report, the U.S. National Academy of Sciences was very cautious on this point, based on the following observations. Yields of approximately 100 q/ha, much higher than the average, have been attained for corn, but only under experimental conditions. U.S. farmers produce yields of only 65 q/ha, which is high for a national average. For wheat the situation is even less favorable, since it has proven difficult to increase the quantity of the leafage per acre given the position of the ears. Many European farmers, however, have produced 100 q/ha. The biological limit has apparently also been reached, within the boundaries of current scientific knowledge, in the quantity of eggs a chicken can produce annually. Another critical subject of research is the most efficient transformation of livestock feed calories into meat calories. The inefficiencies of the current system are obvious. For 100 calories of feed, chickens provide only 12 in return. This meat/feed ratio climbs to 20 percent for pork but drops to 6 percent for cattle. It is both possible and vital to improve these yields, but by how much, no one can say.

For the exporting developed countries, the USDA reasoned that it does not appear prudent to count on a very high growth rate. It can be seen in Table 1.2 that the annual increase in yields was 2.9 percent for wheat and 4.9 percent for corn between 1961–63 and 1971–73. The figures used in 1974 by the USDA's GOL model, on the other hand, are considerably lower, between 1.5 percent and 1.6 percent per year, while the results observed for the period between 1971–73 and 1979–81 are 0.8 percent and 1.8 percent respectively.

The last exogenous factor directly affecting world food supplies is *climate*. Much has been written, including a famous study by the CIA, on two alarming but contradictory theories about climate. The question is no longer whether the world is growing colder or stabilizing but whether it is growing colder or warmer. In this regard, the cautious scientists remind us that there is a disparity between global and local temperature ranges. Climatic cycles stretch over several hundred, indeed thousands of years. Within these cycles, short-term variations occur, but they do not alter the long-term trend.

Reliable meteorological records have been kept for only some 150 years. They are clearly inadequate to discern a long-term cycle. Nor do they help in determining short-term trends, over ten years, for example. During the last century and a half we have seen several of these brief variations, in opposite directions. None of the agricultural studies bases its projections on lasting changes in climate. Only a few mention the potential effect of climatic aberrations. In this sense, a drop of 3 degrees Celsius in the annual average temperature of the USSR and Canada would cause a decrease of 50 percent in wheat production in these two countries. The consequences of this change would be tragic, but the hypothesis is extreme.

Climatologists also note that its latitude places the United States in an advantageous position in either case, whether temperatures rise or fall. If temperatures rise, the Corn Belt would become a new California or Brazil; if they fall, it would replace the Canadian prairies, which would then be unsuitable for cultivation.

What, then, does this examination of exogenous factors tell us about the relatively optimistic studies cited above? The slowdown in world population growth, or at the least a leveling off, appears well established. With respect to supply, the situation is less definite. It is uncertain what role climatic factors will play, and their medium- or long-term influence cannot be foreseen. Only arable land resources and productivity gains remain to be assessed. Land resources are abundant, although unequally distributed. If one considers that only half of all arable land is in use today, a growth rate of 2 percent per year would exhaust this resource in thirty-five years. In one respect, this is not very much time. In another, it is more than adequate. If no other efforts are made, the next generation will feel the constraints of this limitation. Yet there is sufficient time to introduce new technologies, thus making the development of additional land unnecessary. Policies to improve productivity are therefore imperative. Considerable opportunities for progress are open to the developing countries, and technology transfers must play an essential role (see Chapter 6 for the role of the research funds).

In purely physical terms, therefore, the world appears to possess the necessary resources to assure adequate supplies of food, but the more alarming problem of equitably distributing these supplies persists. The hypotheses envisioned concerning population, arable land area, crop yields, and climatic variations have led to a conclusion common to all of the studies: there is a great disparity between the conditions for growth in the industrialized countries and those in the developing countries.

There is a certain harmony to be found in the international marketplace, where the surpluses of the exporting countries are exchanged for goods from the developing countries. This exchange, however, must be made at an equitable price; otherwise the developing countries will remain indefinitely in a state of economic dependence, jeopardizing far more than their economic stability.

Having discussed the exogenous (or technical) assumptions, we will now turn to assumptions which are more economic in nature.

ECONOMIC ASSUMPTIONS: THE NEED FOR TRADE WITH THE THIRD WORLD

The deficit in grain trade for the market economy developing countries has evolved as follows: 18 Mt in 1969–72, 41 Mt in 1977–78, and 51 Mt in 1980–82. These figures do not include the Chinese deficit estimated at 18 Mt in 1980–82 but do include the Argentinian surplus of 14 Mt and the Thai surplus of 4 Mt.[14] Of course, this is an *observed* trade deficit (including 10 Mt of food aid) and not a food deficit *calculated* as the difference between the food needs of the populations and the real resources available to them.[15]

None of the studies shown in Table 1.1 projects a deficit of less than 34 Mt for 1985, a figure which in the USDA study corresponds to the green revolution hypothesis. The University of Iowa report goes so far as to envision import needs of approximately 113 Mt. On the average, 1985 imports are projected by the studies to equal approximately 66 Mt, which, with a margin for error, would mean a range between 40 and 90 Mt. These needs are substantial, representing about 12 percent of grain consumption in the developing countries, including China. The corresponding expenditures could equal between $10 billion and $25 billion. To this must be added imports of meat, dairy products, and various beverages, which urban populations in the developing countries are consuming in increasing quantities. These nongrain imports alone represented $48 billion in 1980–82, as compared with their agriculture, forestry, and fish exports of only $62 billion. Overall, agricultural trade by the developing countries has barely been balanced since 1981.

At the risk of appearing too brief, our discussion has been limited to a few of the most important points, mentioning the many interactions between the agricultural sector and the secondary and tertiary sectors.

Among all the technical assumptions and all the aggregate results of the different studies, *one question is central with respect to future international exchanges of agricultural products: Will the exchanges be commercial or noncommercial?* In other words, what proportion of the flow of agricultural products will pass through the marketplace? The answer to this question is vital because it will determine which of two quite different paths will be taken between now and 2000. The consequences of each option must be carefully examined from both theoretical and practical points of view. Afterward, it will be possible to specify options open to the world's largest importer of agricultural products, the European Economic Community, and the world's largest producer, the United States.

The most important finding is that the sufficiency or even surplus in food supplies, which has been projected for the world as a whole, hides an underlying tension between two regional imbalances: immense food surpluses in the industrialized countries and continued need in the developing countries. None of these studies, it must be recalled, projects self-sufficiency in grain for this latter group. On the contrary, in most cases the deficit is projected to continue increasing for three groups: some countries of Asia (Bangladesh, China), the whole of Africa, and the Near East. For countries in these groups, imported grain is far from a luxury. In all of the studies, projections are based on the minimum level required to assure more or less adequate nutrition (2,335 calories per day for Africa, 2,223 for Asia). These irreducible requirements cannot be satisfied by local production. Imports are thus an absolute necessity.

The scope of these net imports could be traced up to the year 2000 by using the USDA's GOL model results or the International Wheat Council's (IWC) *Long-Term Grain Outlook* forecasts. The production of grain in the developing countries with a market economy (excluding China) is supposed in GOL to grow at an annual rate ranging from 2.8 percent under the protectionist scenario (III) to 3.8 percent under the green revolution scenario (IV). The use of grain in the same countries will be growing at a rate ranging from 3.0 percent under III to 3.7 percent under IV and II (free-trade scenario). The net trade deficit resulting from the extension to the year 2000 of these rates will be as follows:

I (status quo scenario)   = 152 Mt
II (free trade scenario)   = 209 Mt
III (protectionist scenario)   = 146 Mt
IV (green revolution scenario) = 123 Mt

With the same base figures, IWC is forecasting a year 2000 deficit of 137 Mt. The scope of the possible deficit is very large, the IWC projections being in the middle of the status quo (I) and of the green revolution (IV) scenarios.

These deficits are very large, especially if we compare them with what they were in 1980 (68 Mt) and 1960 (17 Mt only). In forty years they could be multiplied by 7 to 12.

Moreover, if we are especially concerned with the low-income developing countries (such as India, Bangladesh, and most of the African sub-Saharan countries), their deficit in the year 2000 as forecast by IWC will reach 61 Mt as compared to only 13 Mt in 1980 and 7 Mt in 1960.

Here, the nature of commercial transactions becomes a critical factor. If they are carried out at market prices, imports must be paid for, with the assumption that developing countries can successfully generate a trade surplus to equilibrate their balance of payments. Such a surplus, in turn, presupposes sustained, vigorous growth in their economies and a free trade environment.

In the inverse hypothesis, commercial exchanges are few or nonexistent. The physical transfer of food from the industrialized countries to the countries in need could take place, but in the form of food assistance, that is, the free supply and possibly transportation of food to the ports of the developing countries. Beyond its destabilizing influence, such assistance is costly to the industrialized countries, and the motivations for providing it must be examined in detail.

The Best Case Scenario: Vigorous Growth in Commercial Transactions; A Deficit in Meat

Under this optimistic hypothesis, almost all trade would take place at market prices. The deficit for the developing countries is not alarming. On the contrary, it simply indicates that they are pursuing efforts to accelerate their growth. In particular, they may be producing industrial goods for export at a relative price favorable enough to permit them to pay for food imports.

The optimism in this scenario is based on a hypothesis of vigorous growth in the developing countries (see Table 1.3). The conditions necessary for such hypothetical growth must be examined carefully to determine if it can indeed be achieved. It may well be that existing structural limitations are too great for the economic systems of these countries to overcome. Although we will not

TABLE 1.3. Best and worst case projections (annual growth rate 1980–90 in percent)

	Reference Period	Best Case	Worst Case
GNP growth rate for developing countries[a]	5.2	5.6/6.6	4.6
Growth rate of manufactured product exports from developing countries[a]	10	11/13	9
Growth rate of per capita income[b]			
Developing countries	2.0	4.9	2.2
Developed countries	3.3	3.8	2.1
Growth rate of agricultural supply[b]			
Developing countries	2.6	3.2/3.8	2.8
Developed countries	2.6	2.6/2.9	1.5
Growth rate of yields[b]			
Developing countries	1.6	2.0/2.9	1.9
Developed countries	2.4	1.5/1.7	1.8
		1985	1985
Grain surplus available for export from developed countries[b]	+31 Mt	+71 Mt	+44 Mt
Grain deficit of developing countries[b]	−21 Mt	−71 Mt	−41 Mt

[a]Source: Robert S. MacNamara, speech given before the Council of Governors, Belgrade, October 2, 1979. The reference period under study is 1970–79.

[b]Source: USDA, Economic Research Service, *The World Food Situation and Prospects to 1985* (Washington, D.C.: USDA, 1973). The reference period under study is 1969–72.

attempt to examine all of these limitations in detail, the following two cases will play a determining role.

The Relative Caloric Contribution of Grain and Meat. Table 1.4 illustrates the extraordinary disparity between the relative caloric contribution of meat in the industrialized countries (31.3 percent) and the developing countries (8.5 percent). The disparity is similar for protein. Eighty percent of the protein consumed in the developing countries is provided by grain and vegetable products, whereas in the industrialized countries, 55 percent is furnished by meat, milk, and eggs. We do not expect that the developing countries will ever match the dietary habits of the industrialized countries. Assuming an overall economic growth rate for these countries of 5.6 percent, however, as projected by the World Bank in its optimistic scenario, it is logical to assert that the relative dietary contribution of meat will be much greater than it is today. A precise projection of the increase in demand for meat and milk is not easy. Transition from the concept of basic nutritional needs to that of actual demand is realized through the mechanism of domestic market prices. Furthermore, the level of consumption is determined by two factors: price and income. When income

TABLE 1.4. Characteristics of per capita food consumption, 1972–1974

	Energy		Protein	
	Calories per Person per Day	Percentage of Animal Origin	Protein Grams per Person per Day	Percentage of Animal Origin
World	2,544	17.3	68.3	34.7
Developed countries	3,371	31.3	97.4	55.3
North America	3,522	35.5	104.1	68.3
Western Europe	3,385	31.7	93.0	55.3
Oceania	3,364	40.7	100.7	66.5
USSR and Eastern Europe	3,457	27.3	102.3	48.5
Developing countries	2,207	8.5	56.3	20.3
Africa	2,111	6.0	52.4	17.9
Latin America	2,535	15.8	64.8	38.6
Middle East	2,439	9.5	67.9	20.2
Far East	2,039	5.6	48.8	15.0
China and Asian socialist countries	2,283	9.1	61.6	19.3

Source: FAO, Production Yearbook, 1977, Tables 97 and 98.

increases and prices remain constant,[16] consumption should increase. This, at least, is the hypothesis for nonluxury food products. What percentage of consumption will be translated into demand for meat when income increases by one unit, however, remains to be determined. The answer is difficult because different socioeconomic classes have different propensities to consume. Thus the increased income effect and the substitution effect must both be measured to evaluate the elasticity of meat consumption precisely in relation to income.

Even approximations are difficult. The range of potential annual growth hypothesized by the USDA is 2.2 to 5 percent, which would result in an increase in income over ten years of between 24 and 63 percent. Such increases are not negligible, and if the maximum increase occurs, a revolution in the consumption habits of the developing countries can be expected. It would be possible to gauge the extent of this revolution if precise figures on food demand elasticity in relation to income were available, but we can be certain of only two general trends: (1) elasticity is greater for low-income than high-income groups, and (2) in all countries, elasticity is greater for meat than for other basic food products.

These trends can be seen in Appendix 3 on the elasticity of food demand in relation to income measured from the standpoint of the agricultural producer—in other words, the demand elasticity with which the producer must contend. Unfortunately, the disparities are great between the elasticity calculated by Gale Johnson of the University of Chicago and that calculated by the FAO. We believe that the FAO figures are probably more realistic. But the problem of estimating elasticity is extremely complicated, particularly for food products. It is necessary, first, to be sure that the specification for the demand factor is correct, and second, that the estimation technique is appropriate, particularly in the frequent case when price is determined by a government rather than by the market.[17]

Lacking an overall study considering these difficulties, we sought to determine which of the studies took the effect of income distribution specifically into account. Unfortunately, the USDA team, the only group to consider the effect of prices, does not enter into the redistribution effects of the growth rate it assumes. The OECD study, however, does attempt to deal with this problem. Its overall and extremely tentative conclusions show the importance of this question. Based on an overall annual economic growth rate of 4 percent for the country under study, the OECD projects a highly unequal progression in food consumption. In particular, if the increased income flows entirely to the richest quintiles, demand for animal products will increase by 0.77 percent per year. If this income is distributed to the three poorest quintiles, this figure will increase to 5.75 percent. Finally, if the new wealth is distributed equally, the growth rate in consumption of animal products will be about 2.9 percent.

These figures are too aggregated to permit definite conclusions, but they are useful for the purpose of estimation. For example, it is easy, using Appendix 3, to perform the following calculations. Taking 2.9 percent as the annual growth rate in animal product consumption, the share of this consumption in the daily diet in the developing countries will increase from 8.5 percent in 1973 to 13.8 percent in 1990. If, on the other hand, we use 5.7 percent as the growth rate, the share of animal product consumption will reach 21.8 percent in 1990. Although this would represent significant progress in relation to the present situation, it would still fall well short of the 1973 level in the industrialized countries (31.3 percent). Of course, it is impossible to anticipate changes in consumer taste. In particular, the demonstrator effect, which causes a very high elasticity in demand for meat in relation to income among middle-income groups in the Western countries (second, third, and fourth quintiles), may not be manifest in the developing countries. It should not, however, be excluded as a possibility. Above all, the great disparity between developing countries and Western countries suggests the possibility of greater tensions than foreseen in the market for meat within the developing countries, if economic growth is indeed sustained there.

This observation is borne out by the rapid pace of urbanization. The MacNamara report before the Belgrade Conference states that 760 million people are now crowded into the tentacular urban centers of the Third World and that their number will double in the next ten to fifteen years.[18] By the end of the century, three-fourths of the population of Latin America and one-third of the population of Asia and Africa will be urbanized. In the year 2000, *forty cities will have more than 5 million inhabitants, whereas in 1950 only one could be found among all the developing countries.* There is a definite link between urbanization, Westernization, and consumption of animal products.

The pressures on meat prices in the developing countries will be further accentuated by the difficulties in producing animal product staples. The case of meat provides an illustration. Increased meat consumption makes a country more vulnerable to instability caused by market disruption, particularly because meat production is sensitive to the internal economic situation. There are two reasons for this vulnerability: the highly elastic demand for meat in relation to income and the existence of cycles, in which price expectations are often "adaptive." Unlike "rational" expectations, adaptive expectations are by definition shortsighted and result in successive adjustments in quantity in relation to price, producing a "cobweb" effect, hence the market instability. Using rational expectations, however, agents will make the optimal adjustments based on the information they possess. The concept of rational expectations is embodied in the futures market. The vitality of many of these future markets, however, does not seem to have altered the behavior of American farmers. Furthermore, it has not been proven that rational expectations are always optimal.

These cycles are well known in a number of countries, France in particular, where successive phases have been observed in the pork and beef cycles. In theory, this increased instability should be compensated by a lower sensitivity to variations in climate, which is great when grains are the preponderant crops. In reality, however, this is not the case, since livestock are often fed with natural proteins during a large part of the year. A poor harvest combined with a decreasing phase in the meat cycle could have disastrous consequences for prices. Another source of instability is exposure to variations in overseas markets. Meat is a difficult product to store. Most meat is consumed in the country where it has been produced. For this reason, international markets are narrow, and small changes in quantity cause large swings in price. Another factor is the dependence on inputs. Several kilograms of grain and soybeans are required to produce 1 kilogram of meat. Grain requires fertilizer, and few countries produce soybeans. In France, for example, according to Bernard Auberger, more than 85 percent of protein-rich livestock feed used in the 1970s was imported.[19] In addition, one of the basic characteristics of the world grain and soybean markets is great price volatility. The monthly average price of wheat, for example, has evolved seasonally from one to four as shown in Figure 1.1. Similar variations occur in the market for soybeans or corn.

All of these uncertainties are constraints on the production of animal products, particularly in countries that do not possess adequate instruments to contend with them. To counteract these factors and increase production, countries have resorted to supporting prices at a level above the market price, either

22

FIGURE 1.1 Price volatility of wheat, 1971–1981 (base 100 in August 1971 = $1.28/bushel)

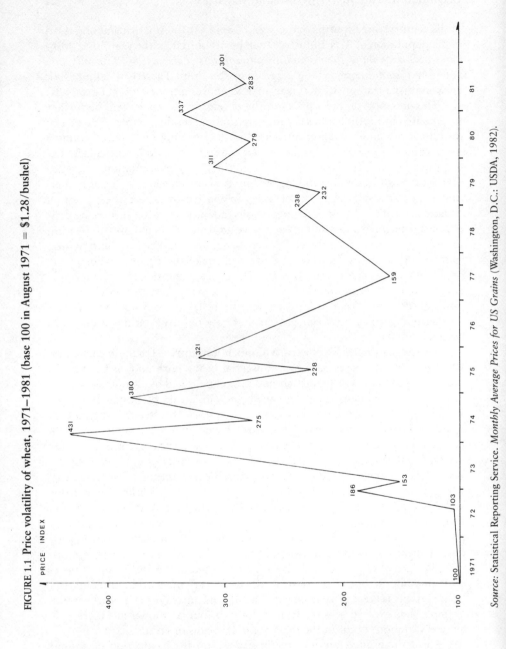

Source: Statistical Reporting Service. *Monthly Average Prices for US Grains* (Washington, D.C.: USDA, 1982).

by abstaining from international competition and trade or by establishing an extremely costly system of subsidies to improve farm income or to compensate the difference with world prices.

In the absence of rapid and fundamental discoveries, acute pressures in the meat markets within the next few years could cast doubt on the earlier optimistic conclusions. To succeed, it will be indispensable to stimulate production and stabilize costs and prices.

The Problem of Industrial Export Markets for Food-Importing Countries. To be able to purchase the necessary commodities to feed their populations, nourish their livestock, and produce agricultural products (fertilizer, pesticides, agricultural equipment), developing countries must acquire export markets for their industrial products. A country can survive without a certain number of manufactured products (cars, refrigerators), but not without food. It is not sufficient to project a high rate of growth for the developing countries; the structure of their overseas trade must also be projected. What products will be traded and at what price? Within the optimistic hypothesis, the developing countries exercise effective control over only part of the necessary conditions. Their fate lies, to a large degree, in the hands of the industrialized countries.

For growth rates in the developing countries to increase from 5.2 percent per year, the rate observed during the 1970s, to 5.6 or 6.6 percent, the rate desired by the president of the World Bank[20] for the period 1980–90, exports of manufactured products from the developing countries must increase by 11 to 13 percent per year, as compared with only 10 percent in the preceding decade. These figures are extremely optimistic. Slowed growth in the industrialized countries and the resulting increase in protectionism call for serious consideration of the protectionist scenario.

The Worst Case Scenario: Self-Sufficiency in the Third World and Europe

This case consists of a rate of growth in the developing countries much slower than expected. This rate could go down to a level much lower than the one projected in Table 1.3 (4.6 percent).

It is reasonable to speculate that insufficient internal demand could in itself constrain supply. The possibility of such saturation, however, is a subject for study which has no place here. Slow growth in the developing countries will therefore be treated as an exogenous assumption.

The worst case scenario postulated by the USDA corresponds to alternative III: protectionism. Under this scenario, the rate of growth in income for the developing countries is only 2.2 percent, compared with 4.9 percent under the free trade scenario. This calls for several observations:

1. A significant decrease in international trade would occur. Total world exports of grain would fall from 193 Mt to 117 Mt. This would be accompanied by a fall in the price of wheat to 1970 levels and, more important, by a severe drop in the price of meat to levels below 1970 prices.

2. The exportable surplus from the industrial countries would drop from 71 Mt to 44 Mt of grain (see Tables 1.1 and 1.3).

3. The effect on developing countries would be the most fascinating. Their trade deficit in food would be *lower* than under the status quo or free trade scenarios. Their net import requirements would drop to 41 Mt, or about 30 Mt less than under the optimistic free trade scenario. The decrease in income would cause a decrease in consumption.

4. Restrictions on food consumption in the developing countries would lend greater urgency to the issue of food assistance. None of the studies discussed here takes this factor into consideration. The volume of 10 Mt of grain recommended by the World Food Conference in 1974 has nearly been attained in the 1980s. Would this level persist in the event of generalized protectionism? The answer is highly uncertain, although the figures show that the surplus in the industrialized countries would decrease less rapidly than would the deficit in the developing countries. Thus it is possible to arrive at a minimum level of food aid of approximately 6 Mt.

The elements of this pessimistic scenario, which will reveal the importance of the analysis further on, are as follows:

1. International commercial trade decreases substantially.

2. A considerable portion (15 to 25 percent) of international trade is carried out on a noncommercial basis in the form of food assistance, especially for the low-income countries.

3. The developing countries find themselves with a reduced deficit, which they cannot pay for entirely.

4. The industrialized countries find themselves with an enormous surplus, for which prices are depressed.

The dynamics of this pessimistic scenario are dominated by two central characteristics: the temptation to retreat into self-sufficiency and the existence of a surplus in the industrialized countries.

Isolationist Tendencies. This tendency refers to a desire that could emerge in certain developing or already developed countries to give priority to agricultural self-sufficiency. If this direction were taken, it would result, as will be demonstrated, in a slowdown in the growth of gross national product (GNP) because of a lower return on capital invested under these conditions. If, on the other hand, industrial production does not increase rapidly enough, or if export markets do not materialize, a certain number of countries would, without question, close their economic borders. The isolationist reaction would be triggered by the lack of outlets for the industrial products of the developing countries.

Through such protection, the governments of the developing countries would attempt to assure a sufficiently stable and adequate income for their farmers to bring about a level of self-sufficiency qualified as underconsumption because only the minimal caloric requirements would be satisfied.

It is necessary, at this stage, to attempt to answer a question which many agronomists and proponents of Third World agricultural development have persistently raised: Is there not a basic incompatibility between industrialization efforts and the indispensable maintenance of a dynamic agricultural sector in the developing countries? Is not economic autonomy the only solution?

This debate is not new. Experts have been arguing about the success of the green revolution for several years. We will not take up all the contentions here, but since "technological" scenarios have raised the possibility of greatly increasing yields, what can the economist conclude?

Above all, it is necessary to examine closely the growth rates being postulated. In the United States between 1948 and 1973, the growth rate in yields per hectare for the agricultural sector was only 1.6 percent per year. Achieving 2 percent, therefore, constitutes a veritable exploit. What were the factors permitting the achievement of this increase? The steady decrease in farm population, increased mechanization, and ever more intensive use of chemical fertilizers would appear to indicate the substitution of variable capital for human labor as the primary factor. If this were indeed the case, such a solution would probably be conceivable in the developing countries, setting aside the cost in human terms. Unfortunately, it is not the case. Granted, this substitution played a role in American agriculture, as will be shown in detail further on. But an examination of the statistical history shows that variable inputs (particularly fertilizer) quickly reach a point of diminishing returns. In fact, the point at which marginal productivity becomes negative (maximum production) is rapidly reached. This occurs largely because of the indivisible nature of using machines: certain input ratios show increasing yields, but this is not the case for fertilizer alone. For example, in the United States the agricultural output per unit of physical variable input *declined* from 113 to 103 (base year [100]: 1969) between 1948 and 1973 (see also Appendix 15). It is just as well, however, that the substitution of physical inputs for labor is not the only means to increase yields; otherwise, the developing countries would rapidly experience budgetary constraints because of the high cost of energy.

What, then, are the fundamental means of improving productivity? They are the more qualitative changes, those affecting machinery (to a certain extent) and soil and labor (to a large extent). In other words, improvement in the quality of machines, education of farmers, and land distribution are the primary factors in the productivity gains experienced in the United States. It would take too much space to list all the studies that have arrived at this conclusion. Generally speaking, if qualitative changes are introduced in the factors of production, considerable decreases are observed in productivity gains. This theory is now well known, but one of the first to apply it to agriculture was Zvi Griliches,[21] who obtained a decrease of 30 percent in the contribution of quantitative factors (labor, soil, machines, fertilizer, and the like) to productivity gains after introduction of the qualitative elements. Furthermore, Appendix 4 clearly shows the relationship that could exist between the average size of American farms and their average yield.[22] In fact, although the extent of improvement in the quality of human inputs, labor in particular, seems to play an important role in the econometric explanation of productivity gains, a certain degree of uncertainty persists when attempts are made to quantify the contribution. Work undertaken by Yair Mundlak and by Lawrence Lau and Pan Yotopoulos testifies to the difficulty of this problem.[23]

In conclusion, *yields are maximized when variable capital (fertilizer, machinery) is applied by a competent farmer over a sufficiently large area.* Furthermore, even though the minimum land area must be defined differently, depending on the relative price of labor and various inputs, and may be completely different in India or Iran, for example, this minimum area is rarely obtained in the developing countries (or for that matter in the industrialized European countries).

The success of the green revolution probably depends above all on the apportionment of land and the education of farmers. Education is costly, requiring both time and instruction facilities as well as highly developed research. Educational prospects, therefore, constitute the first factor tending to dampen enthusiasm over the possibility of attaining productivity gains of 2 percent in the developing countries.

But there is a more important factor. Let us suppose a growth rate in income of 4.9 percent for the Third World (scenario II of the USDA study, which also was apparently used by the University of California). In reality, this figure hides a much higher growth rate for nonagricultural production. Indeed, to attain an annual growth rate of 3.2 percent in agricultural production (2 percent for yields and 1.2 percent for land area), massive investments in the agricultural sector will be necessary. In the case of land reapportionment, for example, there is a minimum land area on which a farmer can satisfactorily increase his productivity, but this minimum area can be obtained only through the purchase of land. Capital is therefore necessary—a rare resource in the developing countries. There is a fringe of the population with savings, often quite extensive, but these people are rarely willing to invest in activities in which the return is uncertain and inferior to what they would obtain by investing overseas or in the secondary or tertiary sectors. To channel the needed financial resources toward agriculture, governments must intervene, either directly or by creating externalities to attract private capital. But at least in the beginning, the return on invested capital will be lower in agriculture than in the rest of the economy. Thus, to attain an overall rate of growth of 4.9 percent, it will be necessary to invest considerably more in agriculture than in industry.[24] Since agricultural production grows more slowly than production in any other sector, vigorous growth and extremely high efficiency must be assumed for the rest of the economy. Finally, the unit cost of production for nonagricultural goods must be maintained at a lower level in the developing countries than in the industrialized countries, so that the real terms of trade do not deteriorate for the country in question. From this viewpoint, it is easier to understand the importance for the developing countries of access to export markets in the industrialized world.

To sum up, *two conditions are necessary for substantial economic improvement in the Third World: markets for their products, either in the domes-*

tic market or in the export market (at prices competitive on the world market), and government policies that make agricultural investments attractive, in particular by guaranteeing a certain stability in agricultural prices and income.

We can conclude, therefore, that the tendency toward self-sufficiency, despite its appeal for governments that wish to assign top priority to agricultural development, may in fact be counterproductive. This realization is evident in China's current policy of the "four modernizations," involving increased recourse to foreign exchange, as well as in India's farm programs promoted by Charan Singh or Rajiv Ghandi to facilitate the purchases of foreign agricultural technology. A similar evolution will likely take place in Iran, which probably will be unable to assure the well-being of its urban population and the investments and training needed by its agricultural sector unless it substantially increases its exports of petroleum.

Investing in agricultural training and directing capital to the agricultural sector are therefore costly policies, for which the return on investment is misperceived by governments in the short and medium term. Only governments, however, are in a position to lead their countries toward an equilibrium in food supplies and demand.

Agricultural Surpluses in the Industrialized Countries. The second premise of the pessimistic scenario is a surplus in the production of the industrialized, exporting countries.

The notion of "surplus" is complicated to explain on an economic level. It will be discussed in detail in Chapter 4. For now, we will simply observe that surpluses exist when supply and demand levels at a given moment are such that the government or some other organization must intervene to support prices by amassing public stocks.

For an industrialized country, which in a "normal" year produces a structural surplus, agricultural policy poses the following dilemma. To simplify, we will assume that there is a minimum size beyond which yields are constant (labor productivity in relation to fertilizer use or land productivity in relation to fertilizer use). If all producers operate farms of approximately this size, they will accept limitations on the area they can cultivate during years of exceptional abundance in exchange for greater price flexibility for the final product during other years. But in reality this assumption implies that during these other years, there will be no surplus. If, on the contrary, a large proportion of the producers operate farms larger than the minimum size, two hypotheses are possible. In the first hypothesis, the government enforces a strict policy of individual production limits in exchange for price guarantees. Over time, such a policy results in farms of a uniform size, optimal with respect to output supported by price guarantees. Once again, therefore, surpluses tend to disappear, following a costly transition phase. In the second hypothesis, the production limit and price support policy are considerably less restrictive; thus, in the absence of climatic constraints, surpluses are not eliminated. In this second hypothesis, inefficient production capacity is the price the country must pay to assure security in food supplies. It can even be imagined that a portion of this surplus would be directed generously toward food assistance. This is a choice for the society as a whole, which, through its national budget, expresses its level of concern for the well-being of disfavored groups at home or in the Third World.

Within this pessimistic scenario, two complementary trends can be distinguished. Out of spite or necessity, the developing countries are increasingly inclined to depend on their own production. This is the self-reliant path. At the same time, however, industrialized countries will be strongly tempted to minimize the cost of agricultural surpluses. In the long term, we must think that a food deficit in the developing countries would pose problems that transcend the economic debate.

The pessimistic scenario leads to a particularly alarming conclusion. Inclined toward autarchy, each part of the world becomes increasingly isolated. For the developing countries, this means a significant slowdown in economic growth and consequently in agricultural demand. The food deficit is reduced but continues to grow over the years. The situation is dramatic for certain regions of the world, which must depend almost exclusively on food assistance. Between 6 Mt and 15 Mt of grain are transferred outside of commercial channels. This aid, however, is the result of large surpluses in the exporting countries. Gradually, these countries grow tired of paying the price to continue providing excess supplies. Policies are established to absorb these surpluses. In the end, food assistance is reduced to a strict minimum.

By the end of the century, it is clear, the world food situation will be characterized by food surpluses in the industrialized exporting countries. This is the nearly unanimous conclusion of all the studies and scenarios developed since 1973. The slowdown in demographic growth in the developing countries and the extraordinary production reserves in the large producing countries revealed by the increase in world prices in 1972 and 1973 have thus completely modified the pessimistic school of thought, which, until 1974, predominated in international agricultural circles.

The unexpected demand in the socialist and OPEC countries is not having the effect previously expected. After a period of upward pressure on prices, which lasted less than three years (1972–75), the generalized increase in agricultural production has more than compensated for this surge in demand.

For the future, however, even in the case of slowed agricultural production in the developing countries, the slow growth in income that would probably accompany this scenario would reduce food demand. It is in this protectionist scenario that surpluses in the exporting industrialized countries would be the greatest. Depressed world prices caused by the relative insolvency of demand would then be the central issue. The governments of exporting countries might oscillate between policies to provide more extensive food assistance to support world prices artificially and policies to absorb the growing cost of surpluses.

These viewpoints, however, overlook the great disparity in local situations. The following two chapters will evaluate the importance of the world's largest importing region, the European Economic Community, and the central role played by the United States among the large exporters. The evolution of these two entities will influence the course of food and agricultural development over the coming decades.

The Agricultural Dilemma
for France and Europe

With 270 million consumers and 5.5 million farms, the European Economic Community holds the distinction of being both the world's largest importer and its second largest exporter of agricultural and food products.[1] Among the ten countries that have constituted the EEC since 1981, West Germany is the world's largest importer, ahead of Japan, the United States, and the United Kingdom; France and the Netherlands are second and third, respectively, behind the United States and ahead of Brazil.

This duality epitomizes the problems of the common agricultural policy, better known by the abbreviation CAP. The results of the CAP have been highly erratic: shortage of powdered milk in 1972, a flood of milk since 1976, an increasing deficit in corn, and growing surpluses in barley and wheat. These are only a few examples, and it is essential to enlighten the reader who thoroughly understands how important Europe is in the world agricultural situation and how much weight its 270 million consumers could represent if they were steadily to increase their purchases on the world market.

The special characteristics of European agricultural production and consumption will be examined first, followed by an analysis of the curious coexistence of surpluses and deficits and of the crucial problem of livestock feed production.

CHARACTERISTICS OF EUROPEAN PRODUCTION
AND CONSUMPTION

Averaging 18 hectares in size, the 5.5 million farms operating in the EEC display decidedly different characteristics from the 2.7 million farms in the United States, which average nearly 180 hectares, ten times the size of their European counterparts.

The traditional European orientation combines mixed farming and animal husbandry. More recently, the pursuit of income levels equal to those in other segments of the population has led a number of the more dynamic European farmers to develop intensive farming operations in French Brittany, the

Netherlands, Belgium, and even Germany, based on indoor breeding of poultry or hogs or on the cultivation of vegetables in greenhouses.

The development of monocultural grain production, which requires large-scale operations, has been possible only in England and the area surrounding Paris. Excluding Greece, the nine EEC nations possess only 37 million ha in the form of larger operations (greater than 50 ha), less than 43 percent of the agricultural land. Among these larger farms, devoted to large-scale grain production, the average size is still well below the 180 ha in the United States: 78 ha in Germany, 81 ha in France, and 130 ha in Italy. Only Great Britain, with an average of 172 ha, approaches the American average.[2]

In any event, nine EEC nations (excluding Greece) possess only 50 million ha of arable land, whereas the United States can cultivate 190 million ha. It should not be surprising, then, that the 28 million ha devoted to grain production in Europe produced only 122 Mt, yet production in the United States totals some 250 Mt, produced on 72 million ha, although populations and standards of living result in similar consumption levels. This physical constraint is the reason behind Europe's difficulty in achieving self-sufficiency in food.

The European Community's dependence on agricultural imports is accentuated by the consumption habits of the average European. There are now 270 million consumers in the ten nations of the EEC, but by the year 2000, after Spain and Portugal are integrated into the organization between 1986 and 1996, this figure should reach more than 330 million.

In 1981, income for the average consumer in the ten EEC nations was $9,070 (per capita gross domestic product), a figure which is approaching the American average of $13,070, particularly in Denmark ($10,360), West Germany ($10,700), Belgium ($9,812), the Netherlands ($9,710), and France ($10,350). But this gap fluctuates rapidly depending on the value of the dollar in relation to the new ECU (European Currency Unit, defined in March 1979 as an average of the currency basket of the EEC countries). Great Britain ($8,515), Italy ($8,290), Greece ($5,200), and Ireland ($5,900) are much poorer.

The typical European, and in particular the typical Frenchman, because of his standard of living and traditions, is a large consumer of meat, particularly red meat, as shown in Table 2.1. France exports its front quarters of beef to Germany and then imports an equal quantity of the more appreciated German hindquarters of beef at twice the price of the front quarters. Furthermore, given the Community's scarcity of grazing land (with the exception of France, Ireland, part of Germany, and Great Britain), it is inevitable that a growing portion of the beef consumed in Europe will be raised on grain and mixed feed. Seven kg of grain are required to produce 1 kg of beef live weight. Pork and poultry require 5 kg and 3 kg of grain respectively per kg of meat (these figures can be reduced by 30 percent if high-nutrient mixed feeds are used).

These dietary habits are therefore costly in grain. Twenty-four Mt of imported American corn would ordinarily be necessary every year to cover the European grain deficit. But, in practice, 16 Mt are now supplied by grain

TABLE 2.1. Per capita consumption of meat, poultry, and fish in 1976 and 1981

Kg/head (carcass weight)	USA	EEC	France
Red meat and offals	83	71	87
Poultry	26	12	15
Fish	8	12	14
Total in 1976	117	95	116
Total in 1981	114	88	111

Source: Eurostat, Agricultural Statistics Yearbook.
Note: The year 1976 corresponds to the average of the EEC with nine member countries whereas for 1981 the average includes the ten member countries of the EEC with Greece.

substitutes such as cassava (manioc) from Thailand (6 Mt) and other substitutes, including bran, molasses, starch, citrus pulp, and corn-gluten-feed (10 Mt), leaving a market of only 6 Mt to 10 Mt per year to be supplied by American corn producers (see Figure 2.1).

At the same time, livestock feed rations (for beef, pork, and poultry) require a protein supplement because corn contains only about 10 percent of vegetable protein, wheat 12 to 14 percent, and cassava only 4 percent. Since their large-scale development in the United States, beginning in 1952, soybeans have been the most economical source of supplemental protein. In 1980, more than 18 Mt in soy meal equivalent of this miraculous product were being imported each year by the EEC, either in the form of oil cakes (residue from crushing beans after extraction of the oil) or of beans, to be crushed in Rotterdam, Anvers, Hamburg, Saint-Nazaire, Brest, and, very recently, Bordeaux. In 1981, imports by the EEC reached 18.7 Mt in soybeans measured in soymeal equivalent, 14.8 Mt in grain substitutes, and 8 Mt in corn. These figures are high for a continent of which the official policy is to attain self-sufficiency in food. They are even more astonishing when viewed in relation to total world exports of livestock feed: approximately 100 Mt of feed grain and approximately 50 Mt of soymeal equivalent in soybeans. The EEC, therefore, uses one-fifth of the world's exportable livestock feed resources in grain and one-third of that in protein.

The United States alone is responsible for about 70 percent of world exports of corn and soybeans. It is therefore understandable why Europe is attempting both to reduce its trade deficit in basic products and to diversify its sources of supply. With respect to grains used for animal feed, diversification has been accomplished in an unexpected and unintentional way: because reduced duties have been granted to cassava imports from developing countries, within a few years, Thailand has succeeded in producing up to 6 Mt of cassava exclusively for hog feed in Europe. This situation has caused discontent among European grain producers, who have called for a ceiling of 5 Mt on European imports of cassava from Thailand.

At the same time, under a new grain policy, the "Silo Plan," the Communi-

FIGURE 2.1 EEC grain and grain substitute supplies, 1970–1980

Millions of metric tons

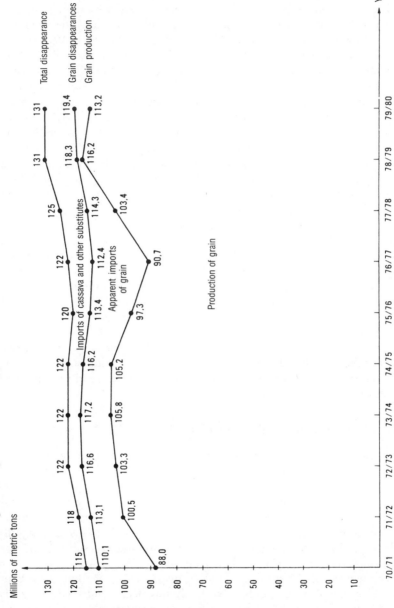

Sources: World Grain Situation Outlook, 1979/80; USDA, *Foreign Agricultural Circular*, February 13, 1980; AGPB, "Le VIIIe plan des céréaliers, importations de substituts des céréales" (Toulouse: French Association of Wheat Growers, June 1980).

33

ty is attempting to promote increased production of corn and wheat for livestock feed (see Appendix 5).

The U.S. monopoly in soybeans has been undercut by the entrance of Brazil and Argentina into this market. Brazil's policy has consisted of taxing exports of soybeans so as to promote the export of oil cakes. Since 1978, nearly two-thirds of the oil cakes imported by the EEC have come from Brazil, and almost all of the unprocessed beans continue to come from U.S. producers, whose foresighted strategy of establishing crushing plants in Europe is paying off handsomely. The production of soybeans in Argentina is increasing steadily and threatens soon to encroach significantly on the U.S.–Brazilian dominance of this market. The growth in world demand for protein-rich nutrients, however, is such that the price for this miracle staple appears unlikely to decline substantially. At best, we can hope to avoid a sudden quadrupling of prices, as occurred in Chicago in 1973, or a recurrence of the following embargo, which lasted several weeks. The memory of this disaster continues to haunt European cooperatives (see Chapter 6 and especially Figure 6.1).

Internally, programs to encourage cultivation of oil/protein–bearing plants have steadily been introduced within the CAP. They are being applied with limited success to soybeans (Community production is only 30,000 metric tons) and with far more substantial results to rapeseed, sunflowers, flax, protein-bearing peas, and field beans. The unambitious objective is to limit European dependence on imports to about 80 percent of its needs. It is not certain that even this level can be achieved, given the extremely rapid growth in consumption.

European food consumption and supplies are therefore difficult to balance in light of the prevailing dietary habits. This imbalance is compounded by the imperfect orientation of its agricultural production, which results in surpluses difficult to sell on world markets.

SURPLUSES AND DEFICITS IN EUROPE

The degree of self-sufficiency in food supplies—that is, the ratio between production and consumption for the EEC both as a nine-member organization and as a twelve-member organization (including Greece, Spain, and Portugal)—will be studied first. Table 2.2, constructed from studies by the Commission of the European Community, shows that the degree of self-sufficiency resulting from the addition of the three countries would not decrease significantly for grains, sugar, and milk and that it would increase substantially for citrus fruits, vegetables, wine, and olive oil.

The paucity of statistical information dispensed by the European Economic Community is regrettable. No study has been conducted on trends toward self-sufficiency within the organization, even though with its enlargement it will be the largest agricultural and food product–importing group in the world. Table 2.2, showing self-sufficiency for 1974–76, underestimates the increase in both consumption and production which would result from the

TABLE 2.2. Self-sufficiency in the European Economic Community as a nine-member and a twelve-member organization: Average production and consumption, 1974–1976

	EEC (9 members)		EEC (12 members)		Deficit to Be Imported (−) or Surplus to Be Exported (+) Mt	
	Quantity Produced (Mt)	RSS (%)[a]	Quantity Produced (Mt)	RSS (%)[a]	EEC 9	EEC 12
Wheat	40.9	103.9	47.9	102.8	+1.5	+1.3
Barley (1972–74)	34.3	101.5	40.1	100.3	+0.5	+0.1
Corn	13.2	54.8	15.9	50.2	−10.9	−15.8
Soybean and oilseeds	[b]	28.0	[b]	[b]	−11.7[c]	−13.6[c]
Sugar (1976)	9.7	104.7	11.0	100.7	+0.4	+0.1
Citrus fruit	2.9	40.1	6.6	73.6	−4.3	−2.4
Other fruit	15.6	78.7	23.0	86.6	−4.2	−3.6
Wine	15.2	99.1	19.8	104.9	−0.1	+0.9
Vegetables	25.6	93.3	[b]	[b]	−1.8	[b]
Milk	98.9	99.1	106.4	98.9	−0.9	−1.2
Beef meat	6.2	94.3	6.8	93.5	−0.4	−0.5
Hog meat	8.3	99.7	9.2	99.4	0.0	−0.1
Sheep meat	0.5	71.4	0.8	78.4	−0.2	−0.2

Sources: Figures from the Commission of the European Community and Eurostat, 1974–77.
[a]RSS = rate of self-sufficiency.
[b]not available.
[c]In tons of SME (soy meal equivalent), 1 SME = 1 ton of soybean × 0.795.

integration of these three southern European countries in the common agricultural policy. It is therefore useful to conduct forecasts based on studies that have extrapolated trends through the 1990s and 2000s for the EEC member nations on the one hand and all of Western Europe on the other. Two studies were released in the beginning of 1984; they have been mentioned in Chapter 1 concerning their results at the world level; they will be used here for the data pertaining to the European continent. One was conducted by Resources for the Future and the other by Winrock International.[3]

These forecasts do not take into account imports of cassava and other grain substitutes, which reached 16 Mt in 1982 (see Figure 2.2). As a livestock feed, cassava substitutes for either European barley or corn imported from the United States. Corn-gluten feed substitutes for Brazilian and American soybeans, American corn, and European barley. The importance of domestic political decisions (CAP reform) has been underestimated by both the Resources for the Future and Winrock studies. The price structure of the Silo Plan (see Appendix 5) will influence the consumption of wheat by livestock, and the quota

36

FIGURE 2.2 EEC grain substitute imports, 1968–1978

Millions of tons

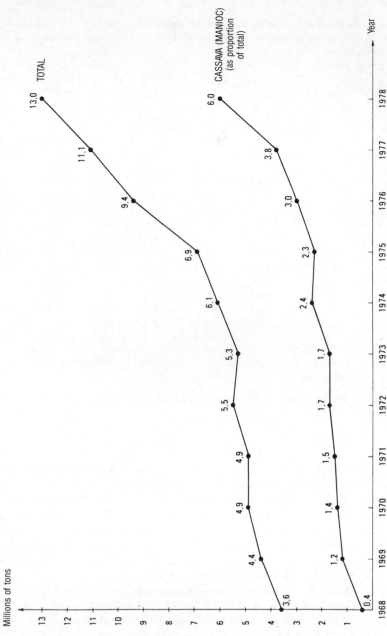

Source: AGPB, "Le VIIIe plan des céréaliers" (Toulouse: French Association of Wheat Growers, June 1980), according to NIMEXE statistics.

imposed on milk or grain production of European farmers will limit EEC imports and exports.

In view of these observations, it is interesting to examine the projections made in 1984 by Resources for the Future and Winrock International. Imports of grain substitutes such as cassava and corn-gluten feed have replaced imports of feed grains, and the EEC (ten-member) is projected to have a 11.7 Mt grain surplus in 1990 or 12.6 in 1993 and 20 Mt in 2000. For Western Europe as a whole, grain supply and demand will be almost balanced in the 1990s, but if Eastern Europe (excluding the Soviet Union) is taken into account, Europe will still be importing 18 to 19 Mt of grain in 1990–93 compared with 28 Mt in 1970 and 25 Mt in 1980. This decrease is largely offset by imports of grain substitutes by the EEC, which none of these studies project for the 1990s but which will probably exceed 20 Mt. It is rather the vegetable protein situation that will steadily deteriorate. According to these studies, the EEC (ten-member) imported 17 Mt of oilseeds (including oil meal as oil seeds equivalent) in 1970, 30 Mt in 1980, and will import 35 Mt in the 1990s and 40 Mt in 2000. For Western Europe as a whole these figures are 20 Mt, 36 Mt, 42 Mt, and 47 Mt in 1970, 1980, 1990, and 2000 respectively. For Eastern and Western Europe they are 22 Mt, 42 Mt, 50 Mt, and 55 Mt.

The analyses of the degree of self-sufficiency or of the net imports and exports of the EEC overlook an important phenomenon: for many products, and particularly for grain, the EEC is both an importer and an exporter. It is interesting that after the creation of the EEC in 1957 and the CAP in 1967, the United States predicted a decrease in agricultural exports to this part of the world. In fact, as shown in Figure 2.3, this did not occur even though many substitutions were made from grains to oilseeds and to grain substitutes as corn-gluten feed. The EEC remains a privileged zone for agricultural imports, but it is also increasingly becoming a permanent exporter of certain products.

An analysis of the preceding supply-demand balance also gives an idea of the hybrid nature of the EEC as both an importer and exporter. For the Community as a whole, the overall deficit breaks down into a surplus in wheat and barley and a growing deficit in corn (or grain substitutes).

In reality, the situation is even more complex because the EEC is an importer of high-quality wheat and an exporter of ordinary wheat, and certain countries can export barley while others must import it. An analysis of the EEC's projected agricultural trade balance shows this situation clearly: Nearly 30 Mt of grain or grain substitute imports are projected during the whole period 1970–90. At the same time, exports, primarily from France, are shown to increase steadily. In 1983, the EEC thus exported nearly 12 Mt of grain, principally wheat and barley, to countries outside of the organization. These quantities could be increased if European imports of grain substitutes continue to grow according to the trend indicated in Figure 2.2.

This analysis may seem excessively detailed, but it is indispensable to show the true nature of the European Economic Community. It is not a self-sufficient

FIGURE 2.3 Food and agricultural exports from the United States to the EEC (nine members), 1957–1979 (billions of dollars) calendar years

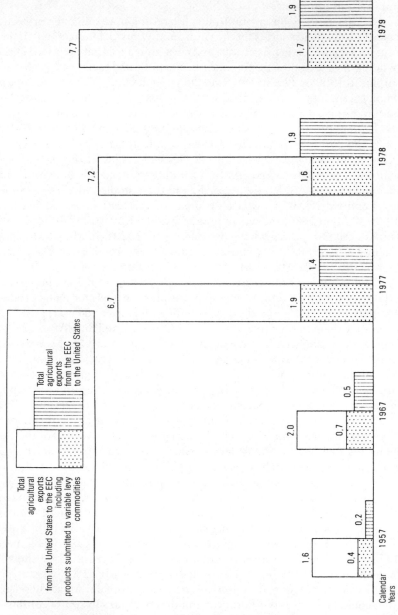

Source: Fatus ERS USDA.

38

zone, but rather a group of countries increasingly exposed to international commerce, despite a very elaborate system of protectionist measures.

Historical consumption habits continue to shape the needs of each of the member countries, and the traditional commercial circuits persist, at least partially, despite the principle of Community preference. The large agricultural countries could thus encounter permanent competition from the EEC for products such as wheat, barley, poultry, apples, wine, milk products, and sugar. France continues to play its traditional role as bridgehead of the Community, although other countries are becoming increasingly important in certain areas of production: West Germany (barley and dairy products); the Netherlands (dairy products, livestock feed, and greenhouse vegetables); Italy (wine, fruits, and vegetables); and Denmark (pork and dairy products). This coexistence of surpluses and deficits indicates that the major concern for this zone is to offset, by whatever means possible (production or importation), its growing needs for livestock feed.

LIVESTOCK NUTRITION IN EUROPE

Is the CAP as it is currently conceived and executed capable of resolving the livestock feed deficit and, by extension, the meat deficit within the EEC?

The first characteristic of livestock nutrition in Europe is the *competition between soybeans and grain*. Indeed, the ratio between the price of soybeans and corn, which in the United States is 3 to 1, is only 1.5 to 1 in the EEC.

In the example illustrated in Table 2.3, based on the figures for May 8, 1979 (date chosen at random), this ratio decreases from 2.59 to 1.19 because of the effect of transportation costs and the levy on corn set on that day at $120.73 per ton.

A calculation for the period 1971–79 was conducted by the USDA comparing the ratio between the price of soymeal (rather than the soybean as in Table 2.3) and the price of corn in Decatur, Illinois, on the one hand, with the same price ratio in Rotterdam following customs clearance, on the other. The mechanism for increasing the price of corn through a levy upon entry into the EEC, described in Table 2.3 for May 8, 1979, leads to the same distortion for the entire period of thirty trimesters examined by Alan Holz.[4]

The ratio between the price of soy cakes and that of corn is on the average

TABLE 2.3. Soybean-corn price ratio for May 8, 1979 ($/metric ton)

	Chicago	Rotterdam Before Levy	Rotterdam After Levy
Soybean	264.00	298.50	298.50
Corn	102.00	131.00	251.73
Price ratio	2.59	2.27	1.19

almost two times greater in the United States than in Europe (from 1971 to 1979, average: 1.86; standard deviation: 0.53). This distortion evidently decreased during the 1972–75 crisis (from July 1973 to September 1975, average: 1.22; standard deviation: 0.12), but it later tended to increase (July 1977 to September 1979, average: 2.48; standard deviation: 0.14). This means that soymeal is increasingly available in Europe at a price equal to that of corn; since March 1974 the price ratio in Europe has almost always been less than 1.20, whereas since October 1976 it has almost always been greater than 2.30 in the United States.

The result is overconsumption of soybeans in Europe. Their share in livestock rations has increased because they contribute both abundant protein and energy. The energy value of soybeans becomes particularly attractive when their price is not too high compared with that of corn. Any examination of the food situation in Europe must therefore take into account both the 25 Mt of imported corn and substitutes and the 15 Mt of soy meal, which at the same time complement and compete with corn.[5]

A second, more recent phenomenon must be taken into account in the study of animal food supplies: *the appearance of cassava and corn-gluten feed as a substitute for corn.*

Increased competition from Thai cassava and American corn-gluten feed has, without question, prevented excessive growth in European demand for corn and thus represents a diversification of EEC supply sources. Because these products have been entering the Community freely, however, they are sometimes priced as low as 30 percent below grains produced within the EEC (see Appendix 5). Negotiations with the governments of Thailand and the United States have been initiated to limit imports of cassava and corn-gluten feed so as to prevent this strain on the CAP from becoming a gaping hole threatening the entire production of grain in Europe, which, after all, amounts to more than 140 Mt.

The case of cassava provides a good illustration of the ambiguity in European grain policy. The livestock feed industry has grown considerably as a result of increased use of balanced nutrients (those containing adequate quantities of energy, protein, and vitamins) for cattle, hogs, and poultry. In 1978 in the nine nations that then formed the EEC, this industry produced 70 Mt of balanced food compounds.[6] Of these 70 Mt, (a) 28 Mt are grains, of which more than half (15 Mt) consists of imported American corn; (b) 13 Mt consist of cassava from Thailand (6 Mt) and equivalent imported products (7 Mt of by-products, molasses, and pulp), which substitute for and have the same energy content as corn; (c) 15 Mt are soy meal or other oil meals such as peanuts; and (d) 14 Mt are various products purchased at the best possible price in Europe from food industries selling their by-products (brewery draft, for example). The European livestock feed industry, therefore, is dependent on external sources for 60 percent of its activity. This is the essential problem for the future of the CAP.

The role of the European Community in the world food and agricultural situation has become fundamental for two reasons:

1. The dynamism of the European economy has resulted in a rapid evolution in lifestyles, particularly in the dietary habits of its population (330 million in 2000), whose consumption is constantly increasing under the effect of successive enlargements. As a result, the EEC is now the world's largest importer of food and agricultural products.

2. Because of its traditions and large rural population, the enlarged European Community is a permanent exporter of certain products (such as milk, wheat, barley, white sugar, wine, and olive oil). It must find markets for these products in competition with the large exporters of food and agricultural products.

It is therefore not by accident that the commercial negotiations held within the General Agreement on Tariff and Trade (GATT), located in Geneva, are concentrating primarily on the dialogue between the United States and the EEC and are dealing essentially with agricultural issues. This dialogue between the world's largest agricultural producer and the world's largest importer of food is of vital importance and will increase as population growth and improved standards of living in Europe drive consumption levels higher.

U.S. ambitions in the area of food and agriculture, which we will now examine, are thus in permanent conflict with those of this resistance side that EEC represents.

The Dominant Role of the United States in the World Agricultural Economy

Many Europeans think of agriculture only in terms of the European Community. This is akin to discussing petroleum without mentioning the Persian Gulf. In this chapter we would like to demonstrate that today the strength of American agriculture places the United States in a position similar to that of OPEC in the field of energy.

This notion is not widely held, nor is it apparent at first glance. There is obviously a gap between the reality of America's formidable agricultural potential and the extent of its actual participation in world markets. In statistical terms, U.S. dominance has been a matter of record since the late 1940s, but for twenty-five years (until 1965) this fact was obscured by a policy of limited agricultural exports.

This chapter will examine the central role now played by the United States in world markets. The first section will provide a few indications of U.S. agricultural potential. Here, the principal question is the following: Using all of its productive capacities, how much could the United States produce in agricultural staples? And within this hypothesis, what would be the U.S. position in world markets?

It is pointless to undertake this exercise without taking the rest of the world into account. The second section will therefore evaluate the agricultural potential of America's principal competitors.

Finally, we will defend the hypothesis that agricultural exports play a role which is so important to the United States that they act as a constraint. As the central pivot in the international system of agricultural trade, the farm sector of the United States depends vitally on increased trade. This, in any case, is the conclusion of the principal projections and simulations made by American authorities.

THE ENORMOUS POTENTIAL OF AMERICAN AGRICULTURE

An Exceptional Presence in World Markets

There is no single analogy apt to describe the place of the United States in the world agricultural situation. We will return once again to the comparison with the Middle East. With respect to petroleum, the extent of our energy dependence is not solely because one-third of the world's production of petroleum comes from the Middle East. It is accentuated by a fact which in many ways is frightening: more than three-fourths of the world's known reserves are located in this region of the world. The control exercised by a small group over our economic future is a matter of great concern.

Is there cause for greater optimism in the area of agriculture? Statistically, U.S. dominance appears less overwhelming. In 1981,[1] the United States produced only (a) 17 percent of the world's production of wheat—more than double that of Canada and France, the second and third largest producers, respectively, in the Western world; (b) 16 percent of the world's production of barley; (c) 17 percent of the world's production of tobacco, almost twice that of the second largest producer, India; (d) 46 percent of the world's production of cotton, excluding the Communist countries; (e) 63 percent of the world's production of corn (the same year the European Community produced less than one-twentieth); and (f) 63 percent of the world's production of soybeans, a proportion that has been growing in recent years, despite the emergence of Brazil and Argentina.

Except for corn and soybeans, these figures do not appear especially dramatic. Nonetheless, they reflect a formidable dominance because they could be increased, even doubled if necessary. Recent calculations by C. Yeh, Luther Tweeten, and Leroy Quance have shown that by using all available arable land, the United States could feed 4 billion people more than 3,000 calories each per day.[2] In other words, the United States could feed the entire world's population and provide a higher-energy diet than currently available (on the average 2,400 calories per day). We will return to a review of this calculation and the validity of its basic assumption further on. For now, we will treat this calculation as an approximation which only the United States could possibly achieve.

Historical Perspective

We can also provide a more dynamic picture of the role of American agriculture in world markets. To do so, it is necessary to distinguish between the various products. Indeed, as we will see, there is a fundamental difference between a product such as wheat, for which U.S. strength derives from its surplus stock, and products such as corn and soybeans, of which the United States produces more than one-half of the world's total harvest.

With respect to wheat, the U.S. percentage of the world market (13.1 percent in 1983–84–85) is substantially lower than it was in 1949 (18.8 percent), but U.S. dominance clearly does not stem from its production levels. Instead, one need merely refer to U.S. stocks of wheat, which have never constituted less than 30 percent of the stocks of the world's principal exporting countries. Between 1949 and 1985 the average was 50 percent. American wheat producers, therefore, have a double advantage: the capacity to increase production greatly if necessary and the ability to influence world markets decisively through their considerable stocks.

With respect to corn and soybeans, the situation is even clearer. Here, irrespective of stocks, the production figures speak for themselves. From 1949 to 1985 the United States produced an average of 47 percent of the world production; the other major world corn producers are China (15 percent), EEC (7 percent), Argentina (2 percent), Thailand (1 percent); for soybeans the U.S. production has averaged 58 percent of the world total from 1949 to 1985. In both stocks and production, American dominance in these three agricultural products, which represent more than one-fourth of all world exports, is impressive.[3] It is therefore not surprising that the U.S. share of world exports is enormous. The U.S. share of the world wheat exports is almost as great today (35.8 percent for the average 1984–85–86) than it was in 1949 (37.8). For corn, the United States has exported more than 70 percent of the world exports since 1980, as compared to around 35 percent in the 1950s. For the total feed grains the U.S. share of world exports was 59.3 percent for the average 1984–85–86. The same is true for the U.S. share of the soybean and soymeal market, which appears to have stabilized at around 70 percent (see Chapter 6, Table 6.6).

We are therefore faced with an apparent contradiction. As early as 1949, in terms of production, the United States was an unrivaled agricultural power—more so perhaps than today. Yet its presence in world markets and its capacity to influence international trade appear far greater now than in 1949.

This situation is less paradoxical than it appears. To understand the evolution of the American agricultural strategy, it is necessary to refer to the internal agricultural market situation. The American agricultural sector did not exert pressure on the world market during the 1950s as a matter of deliberate policy, dictated by the decision to absorb excess supplies. Indeed, the pendulum swung between a laissez-faire policy, allowing American prices to follow world prices, on the one hand, and the more rigorous interventionist policy, on the other. The second approach was taken until 1964, as we will discuss in Chapter 4. For many years, the United States practiced an isolationist policy, in which exports played a secondary if not negligible role. Nevertheless, the formidable agricultural strength that was to become apparent in the early 1970s was forged during this period.

Absorb Surpluses or Stabilize Production Potential?

For twelve years, from 1952 to 1964, American agriculture was dominated by the problem of "surpluses." There was an excess of wheat, and absorption of excess supplies posed a social and human dilemma. Although they were the source of problems, these surpluses were the result of extraordinary efficiency. A combination of numerous factors (agricultural policy, financial markets, technology, education), which will be examined in successive chapters, made American agriculture one of the world's most dynamic economic sectors. Today, even as broad segments of American industry are being surpassed by foreign competitors (Japan, Southeast Asia, occasionally Europe) American agriculture continues to enjoy virtually uncontested vitality.

To evaluate the efficiency of American agriculture, it is necessary to distinguish the notion of surpluses or production excesses from the notion of production potential. A market can become unbalanced because supply exceeds demand. This condition, which is difficult to define in economic terms, shows an economic sector's capacity to produce more than demand can absorb at a given time. The concept of production potential is different. It refers to the maximum production potential of an economic sector given a balanced market. In the case of a free market, they are determined simultaneously.

Normally, the strength of a national economic sector is discussed within the context of a free market. Agriculture in the United States, however, has been characterized by government intervention. Historically, the debate over agricultural policy until 1960 pertained to means of absorbing excess supplies. It is therefore natural to ask what, in economic terms, is a surplus. A surplus can be defined as two possible situations. The first is when, at a price that "fairly" compensates production inputs, supply exceeds demand. The second, identical to the first, occurs when, at prices necessary to balance supply and demand, insufficient compensation for production inputs is accepted. In both cases, supplies exceed demand. Based on these definitions it is clear that a surplus cannot exist within a free competitive market because of the producer's freedom to enter or leave the market. In other words, when the price falls, certain producers will halt production (or leave the market); when prices rise, new production units appear in the marketplace. These results will be the same regardless of the yields at the individual production unit level. Monopolistic equilibrium exists if yields are increasing, but all production disappears if yields are decreasing uniformly. From this standpoint, agricultural markets display characteristics that prevent the elimination of surpluses because producers are not free to enter and leave the marketplace.

Furthermore, practically none of the other conditions for a competitive balance are fulfilled in the agricultural sector. For example, information on the quality and price of inputs does not circulate with the required fluidity. The cost of the production inputs, such as machinery and particularly credit, is by no means the same for all farmers because of the varied degree of risk each farmer represents for the creditor. Changes in production quantities in response to price fluctuations are not instantaneous. Producers cannot reduce the quantity of inputs to reduce their costs when prices for the final product fall; such flexibility even in the middle term exists only for fertilizers, and even these must be used once they are bought. Because of this series of uncertainties, the farmer is often forced to accept compensation for production inputs, in particular for his own labor, which falls below the level of marginal productivity.

Indeed, some farmers, in effect, accept negative compensation for their work when prices fall, whereas under normal economic circumstances they would halt production. This observation also explains, in part, why farmers choose to maximize production as opposed to profit. When their costs increase, and prices for their products stagnate, they attempt to compensate by increasing production. From a macroeconomic standpoint, this economically irrational (in theory) behavior leads to the appearance of a surplus, as defined above.

If the normal market price, without intervention, is considered too low, pressure is exerted on the government to guarantee a decent income level for *all* farmers—that is, a floor price for agricultural products. This support policy stems from a concern for fairness. But if it is not accompanied by structural measures, it will very soon constitute an unbearable cost for the society as a whole (see Figure 4.2, Chapter 4) because increases in yield result in increasingly greater expense to store excess production.

Such surpluses often took on considerable proportions in the United States during the 1950s. One indication of this situation is the level of stocks amassed by the government. In early 1949, the American government, through the Commodity Credit Corporation (CCC), the official intervening agency, possessed 80 percent of the stocks of wheat. By the end of 1949, this proportion had increased to 85 percent. Furthermore, these "official" stocks represented 33 percent of annual production. If private stocks are added, 39 percent of the total American production of wheat in 1949 remained unsold for lack of domestic or foreign purchasers. In 1957, 110 percent of that year's production was stored in the silos of the CCC (105 percent) or individual farmers (5 percent). That year, the government had more than a full year's production in reserve. In 1964, Luther Tweeten and F. Tyner tried to calculate the "real surplus" in production for the period 1955–61. Later on, Quance and Tweeten applied the same methodology to the period 1962–69.[4] They measured the surplus by adding (a) purchases by the Commodity Credit Corporation, that is, government stocks; (b) the production potential of land voluntarily idled under the support programs; (c) the potential production of land devoted to other crops because of various constraints (individual production quotas, for example); and (d) part of the exports made through the government either in the form of food aid or subsidies. The total value of these components is compared with the potential production to obtain the excess capacity for each year. This approach is consistent with the notion of surplus as defined above.

Results of the Production Surplus Calculation Method

Tweeten and Tyner concluded that for the period 1955–61, the surplus had varied between a minimum of 5.7 percent of production in 1957 and a maximum of 11.2 percent in 1959, or an average of slightly over 8 percent. For the period 1963–69, Quance and Tweeten calculated a minimum of 4.5 percent in 1968 and a maximum of 8.2 percent in 1963. The average in the 1970s was closer to 6 percent, which is logical given the policies followed by agricultural authorities from 1960 to balance production with demand.

These figures call for two comments. The first concerns their effect on prices and farm income. Assuming a demand price elasticity of −.3 between 1955 and 1961, prices would have to fall 24 percent to eliminate the surplus. This figure is 20 percent after 1963. This considerable drop in income for farmers explains the unanimity of their demand for support measures.

The second comment concerns technological potential. Slightly more than 6 percent of total production may seem insignificant in comparison with what we have observed in the wheat and corn markets. But it must be recalled that this constitutes an *overall* surplus for all of American agriculture. In this context, Quance and Tweeten provide a few possible scenarios through the year 1980. Their forecasts assume that technological progress causes no change in the supply curve. Consequently, they are reasoning along the same supply curve. Their work shows that American agriculture, which in 1969 produced a relative surplus of 7.9 percent, could have increased its real production by approximately 8 percent. Their figure is based on the assumption of continued support programs. This means that, according to Quance and Tweeten, the United States has a production potential considerably greater than its real output in 1969.

The pressures American farmers could exert on international markets are therefore considerable, and the temptation is great to compare them to those exerted by OPEC on the petroleum markets.

This picture of American agricultural potential leads to two series of questions. The first concerns its limits, reached or not. Will the American share of world production decrease or increase in the years ahead? The second series of questions relates to the maximum share of the United States in world exports. It cannot be contended that the United States has the same image of dominance which in the past has been projected by the oil-producing countries. Until recently, the United States has shown more discretion, even though since 1973 its weight has become apparent. This discretion was the result of the extremely ambiguous nature of its exports during the last thirty years. Initially considered as a residual to be disposed of, exports have now assumed a much more important role and constitute one of the pillars of the new American agricultural policy. The rest of this chapter will address two questions: What are the limits of American agricultural potential? What is the maximum share of world exports attainable by the United States?

The Limits of American Agricultural Potential

U.S. agricultural strength was forged before World War II, when the first structural surpluses appeared. At the time, these surpluses were considered a burden rather than a benefit. Their existence, however, resulted in the establishment of a series of policies designed to absorb them. In 1959, 22 million acres (9 million hectares) were voluntarily idled as a result of agricultural policy. In 1972 this figure was 62 million acres (25 million hectares). In 1974, after restrictions on area that could be planted were lifted following record exports in 1973, idled land area fell to less than 1 million hectares. In other words, more than 60 million acres (24 million hectares) were activated in less than two seasons.

It was natural, then, for economists to raise the following question: If demand were to grow and create a sharp increase in prices, how much farmland could be activated to satisfy this demand? The methodology appropriate to respond to this question is not easy to define. In an article published in 1975, two agricultural economists with the USDA attempted to collect and evaluate the different procedures used to measure production capacity.[5]

The theory is relatively simple. As often is the case, it is the actual measurement which is difficult. The existence of excess supplies and the notion of production potential are, of course, quite different. In the first instance, demand depends on the market price, among other factors. Supply also depends on the market price, but not immediately, because price is an exogenous factor: it often is the result of political decisions. As a result, since demand, supply, and price are exogenous, only stocks are determined by the model.

Stocks are a measurement of surpluses, but by no means of production potential. Indeed, supply is defined by the producers' reaction to exogenous demand (from the government). Furthermore, farmers must operate under a number of constraints on the amount of land they can use and on the quantities they can produce.[6] Without these constraints, supply would be much greater. This constitutes a fundamental difference from the notion of production potential, which determines the percentage of capacity used at a given free market price. Theoretically, under this hypothesis, supply is by definition equal to demand, and price is determined by the value that assures this equilibrium. The supply function is the aggregate of the production of each producer. Each producer is assumed to respond optimally to economic conditions: each produces to the extent that his marginal income (price) is equal to his marginal cost. This covers the hypothesis of costs, which increase in direct proportion to production increases.

Maximum production capacity can thus be defined as the quantity of production in excess of which overall supply is totally inelastic. Regardless of the extent to which increased demand forces prices upward, production cannot increase. The incapacity to increase production occurs because even for the most efficient producer, the cost of producing an additional unit is greater than the price for that unit. This reasoning takes into account the increasingly inferior quality of new cropland.

The formulation of a supply function that corresponds to the above definition poses a delicate problem.[7] Despite the difficulty of this task, an extremely significant study was undertaken by Yeh, Tweeten, and Quance.[8] The methodology followed by these authors is both simple and rigorous. They begin by defining the external environment in terms of exogenous variables, such as population growth rate, increases in income, and changes in yield. Based on these factors, the authors calculate a level of equilibrium—that is, the price at which supply equals total demand. Next, they factor in alternative prices to formulate a potential supply curve, which gives the excess of supply or surplus for each price level. To this they add the "production capacity"—that is, the production on land that has not yet been cultivated but could be. It is therefore natural that at the equilibrium price, surpluses would be nonexistent but unused production capacity would be pushed to its maximum. When the price increases, unused production capacity decreases and excess supply increases.

The potential production reserve is then defined as the sum, at a given price, of "excess supply" and "unused production capacity."

We will now examine what is known as the "basic" scenario.[9] It is based on the hypothesis that expenses incurred for the improvement of yields and soil will increase at the historical rate of 3 percent per year; the index of prices paid by farmers for nonagricultural inputs will increase by 4 percent per year; and total demand for agricultural products (national and foreign) will grow at a rate of 1.5 percent per year. The ratio of the index of prices paid to farmers to the index of prices paid by farmers equals 100 in 1967.

Under these circumstances, the market is balanced with a ratio of 101 in 1985. Output is 114 percent of what it was in 1967. Excess supply naturally equals zero, and unused production capacity equals 67.5 percent of the production in that year. Conversely, if the ratio increases to 180, supply will be totally inelastic, production capacity will be exhausted, and excess supply will equal 48.7 percent of production (which would then equal 190.9 percent of 1967 production). Table 3.1 summarizes the projections by Yeh, Tweeten, and Quance through 1985. Production is measured as a percentage of what it was in 1967; excess supply and unused production capacity are measured as a

TABLE 3.1. Potential production reserve: 1985 projections for the United States

Scenarios	Price Ratios[a] 100 = 1967	1985 Output 100 = 1967	Excess Supply as a Percentage of Output	Unused Production Capacity as a Percentage of Output	Potential Production Reserve as a Percentage of Output
Status quo	101	114	0	67.5	67.5
	180	190.9	48.7	0	48.7
Acceleration of technological progress	99	114.3	0	68.5	68.5
	180	192.6	49.2	0	49.2
High inflation	92	107.1	0	78.2	78.2
	180	190.9	52.8	0	52.8
High demand	108	119.8	0	59.3	59.3
	180	190.9	45.2	0	45.2
Slow growth in technological progress	112	111.4	0	54.6	54.6
	180	172.2	43.0	0	43.0

Source: C. Yeh, L. Tweeten, and L. Quance, "U.S. Agricultural Production Capacity," *American Journal of Agricultural Economics* 59 (February 1977): 37–48, note 2.
[a]Relationship of the index of prices collected by American farmers at the time their products were sold to the index of prices for their inputs at the time they purchased them.

TABLE 3.2. Unused production capacity according to output prices

Products and prices[a]	Price Ratio Hypothesis[b] 1967 = 100	Land Area Cultivated (in 10⁶ acres)[c]	1985 Output as Compared to 1974 = 100	Excess Supply as a Percentage of Output	Unused Production Capacity as a Percentage of Output
Corn					
(1) 2.52	101	54.4	130	0	36.9
(2) 2.95	106	61.9	100	–	–
(3) 4.94	180	86.3	206	36.9	0
Wheat					
(1) 4.51	101	57.4	124	0	36.9
(2) 4.09	106	53.9	100	–	–
(3) 11.54	180	91.0	196	36.9	0
Soybeans					
(1) 7.39	101	50.8	108	0	36.9
(2) 6.69	106	55.8	100	–	–
(3) 14.44	180	80.6	131	36.9	0
Meat (beef and veal)					
(1) 53.00	101	–	102	0	36.9
(2) 36.00	106	–	100	–	–
(3) 103.00	180	–	163	36.9	0

Source: C. Yeh, L. Tweeten, and L. Quance, "U.S. Agricultural Production Capacity," *American Journal of Agricultural Economics* 59 (February 1977): 37–48.
[a]Price in dollars per bushel or for meat in dollars per 100 pounds.
[b](See note [a], Table 3.1). 100 = 1967.
[c]One acre = 0.4 hectares.
(1) Supply-demand equilibrium price.
(2) Average real price recorded in 1974.
(3) Maximum price at which the entire land area would be used.

percentage of projected production for 1985. It can be seen in this table that in practically every case in which equilibrium is assured, the potential production reserve climbs to 60 percent, reaching 80 percent in one case. This means that under these circumstances, American agriculture could potentially produce at least 60 percent more than it supplies when supply and demand are balanced. This conclusion in itself is impressive. It is even more impressive when individual products are isolated, as in Table 3.2.

A number of remarks are appropriate concerning both the construction and conclusions of Table 3.2. The procedure used by these authors is clear. First, they calculated the price of individual products. It can be seen that the price at which supply equals demand in 1985 (hypothesis I) is, in general, slightly higher than the price in 1974, with the exception of that for corn, which is considerably lower than what it was in 1974 ($2.51 per bushel as compared with $2.95). This is surprising, inasmuch as the inverse is true for the other products. But it must be recalled that 1974 was an exceptional year in which prices for agricultural products were considerably higher than those observed before and after the crisis of 1972–75. Prices for wheat in 1974 reached 313 percent of the 1969–71 level and 153 percent of the 1976–78 level. The same is true for corn, for which the price in 1974 represented 253 percent of the level during the 1969–71 period and 144 percent of the level during the 1976–78 period. The same calculation leads to figures of 242 and 103 percent for soybeans and 130 and 92 percent for beef. It appears, there-fore, that in 1974 there was already a substantial reduction in unused capacity. But it was easier in 1974 to reactivate voluntarily idled corn land than wheat land. Reduction in unused production capacity was greater for corn, which was not the case for wheat and soybeans, for which supply and demand were nearly equal in 1974. This was far from true for beef.

Having made this observation, however, it must be added that the notion of supply-demand equilibrium, as defined in Table 3.2, represents a situation in which demand is already great in relation to supply capacities, with land area constant. The authors then "disaggregated" for each of the four products the total excess supply in their model, which explains why for each product excess supply equals approximately 37 percent of production. It should also be noted that the concept of excess supply does not take into account the decrease in demand caused by a price higher than that which would balance supply and demand. This explains why reserve capacity and excess supply are the same in the case of supply-demand equilibrium as in the case of maximum production potential.

The American farmer still has a large margin for expansion. According to the article by Yeh, Tweeten, and Quance, American reserves of potential cropland alone are enormous. The authors based their projections on an area of cropland in production of 472 million acres (190 million hectares). Accord-ing to M. L. Cotner, M. O. Skold, and O. Krause, reserves that could be converted into cropland amount to 265 million acres (107 million hectares).[10] Of this reserve, 36 percent or 38 million hectares could easily be converted into cropland. In other words, conversion would not require an investment greater than that required for land currently under cultivation, and yield would be comparable. This easily convertible reserve could be operational in one or at the most two years. The only limitation is that, to a large extent, these lands are located in the Great Plains, where only wheat and low yields of barley can be cultivated. These, then, are the data on which Yeh, Tweeten, and Quance base their assertion that if all usable agricultural land in the United States were cultivated, this country could feed 4 billion people a diet of 3,000 calories per day. In other words, this country could satisfy the needs of the current world population, offering a better diet than currently available, since the average caloric intake available in 1966–68 was 2,457 calories per day and in 1975–77 was 2,590 calories per day.[11] This calculation is, of course, purely theoretical, since its realization would require the use of all arable land, including that which is difficult to convert. The cost of the investments necessary to develop this latter category would be considerable.

It is not necessary, however, to envision such an extreme situation. U.S. food equilibrium could be realized in 1985 without resorting to supplementary cropland, and the ratio of prices to costs could increase to 120 without resorting to new cropland. If this ratio were to increase to 180 it would be necessary to use all easily convertible land plus slightly less than half of the land that is moderately difficult to convert.

The limitations of these calculations are obvious. First, there is a limitation inherent to the method used. P. Crosson, a researcher with Resources for the Future, wrote an elaborate criticism of the article by Yeh, Tweeten, and Quance.[12] His criticism does not appear to be completely justified, however, in that its central thesis is based on an assumption of linear supply and demand. Nowhere in the original article is it indicated that this assumption is used. This does not mean that reasonable doubts concerning the supply function used in their analysis should be excluded. This function is based on the NIRAP simulation system, used by the Economic Research Service of the USDA. It has not been possible to gain access to this system, so firsthand judgment of its validity cannot be made.

Other reasons for skepticism stem from the method of calculating the area of land reserves. In this respect, the French agricultural attaché in the United States, Jean-Claude Trunel, in his agricultural report for December 1979, pointed out that the USDA had revised downward its estimates of agricultural land reserves. Instead of 107 million hectares, 38 million of top quality, only 40 million hectares, 12 million of top quality, could be activated without investments. Indeed, according to the analysis published in 1980 by USDA under the title *1977 National Resource Inventory*, 16 million hectares could be rapidly activated, of which 3 million hectares consisted of forest land to be cleared, 4 million hectares of prairies to be drained, and 9 million hectares of prairies to be plowed. But these reevaluations do not change the conclusions indicated before, as long as the ratio does not surpass approximately 140, in excess of which inferior quality land would be used. These reevaluations, however, are cause for some concern, since at the same time it can be observed that arable land area in the United States is diminishing every year by 2 million hectares, or 1 percent, because of urbanization. Projects to strengthen control over the use of agricultural land by nonfarmers were consequently prepared by the Carter administration and implemented by the governments of some states.

In the final analysis, despite these last indications, the reserve capacity of American agriculture is still enormous, and a rare set of circumstances would be required to push use of this land to its limit. What does this mean for world markets? What is the effective power of the United States? The following purely theoretical exercise will attempt to answer these questions by estimating U.S. capacities.

TABLE 3.3 Actual and potential share of the United States in world production and exports

	Share in World Production (%)			Share in World Exports (%)		
	Real 1974 (106)	Potential under Hypothesis 120[a]	Potential under Hypothesis 180[b]	Real 1974 (106)	Potential under Hypothesis 120[a]	Potential under Hypothesis 180[b]
Wheat	14	18	24	50	52	65
Corn	40	49	59	60	81	89
Soybeans	58	67	74	81	90	93

Sources: FAO, *Production Yearbook, 1977;* FAO, *Trade Yearbook, 1977;* C. Yeh, L. Tweeten, and L. Quance, "U.S. Agricultural Production Capacity," *American Journal of Agricultural Economics* 59 (February 1977): 37–48.

[a]The relationship of prices charged to prices paid defined in note [a] of Table 3.1 is 101 at the supply-demand equilibrium in 1985 and could go up to 120 without the cultivation of new lands.

[b]At 180, the supply is inelastic and a part of the lower-quality lands had to be prepared for cultivation.

The Maximum U.S. Share of World Exports

Table 3.3 provides a summary of this discussion and shows estimated quantities. A few observations are necessary on the conception of this table and the figures it contains. The base year 1974 was used to determine the real U.S. share of world markets in wheat, corn, and soybeans. Next, the percentages were recalculated according to two other price hypotheses. These hypotheses were those used by Yeh, Tweeten, and Quance: an increase in the price/cost ratio for farmers from 106 to 120 or 180. Because the United States that year was a net exporter of the three products in question, we added the surplus in American production to world and U.S. exports. In this way, we obtained the new percentages of the world market.

This exercise is only as valid as its underlying hypotheses. Those contained in the Yeh, Tweeten, and Quance model have already been discussed, but the addition of the entire surplus to exports could trigger violent price fluctuations on world markets. These results must be considered instead as approximations, particularly for the 180 hypothesis, which represents the use of the entire "potential production reserve." The 120 hypothesis is much more realistic, however, and in itself highly threatening: it shows a much greater U.S. presence in world grain and soybean markets in all three products.

In reality, this exercise emphatically demonstrates that the United States in very little time could assume a position of quasi-monopoly in world corn and soybean markets. In the case of wheat, the U.S. position appears less advantageous, but this must be compared with the position of its principal competitors. By providing 52 percent of world exports (the 120 hypothesis) instead of 40 percent in 1982, American producers would leave their rivals far behind.

Canada would assume 19 percent of world exports, Australia, 10 percent, and Argentina, 4 percent. Furthermore, these competitors would react much more slowly to the stimulus of higher prices and would be less resistant to a price collapse.

The methodological limitations of the preceding exercise could cause the reader to be skeptical about the validity of the results. A review of wheat stock figures, however, should dispel this skepticism. For thirty years, from 1949 to 1980, the United States never (with the exceptions of 1973 and 1974) possessed less than 30 percent of the stocks of the seven principal exporting countries (Argentina, Australia, Canada, the EEC, Spain, and Sweden).[13] The average is closer to 50 percent. This means that each year, for the last thirty years, the United States possessed almost half of the world reserves in stock, a means of considerable pressure on world prices.

NEARLY UNRIVALED LEADERSHIP

It can be stated with confidence, then, that the United States could assume a position of quasi-monopoly, controlling more than two-thirds of the world's exports of the principal food and feed staples. It is appropriate, however, to examine the production and export capacities of the other "great" agricultural powers.

It is easy to identify the world's great agricultural producers. The annual report on agricultural production by the FAO indicates the countries possessing the most extensive areas of arable land (Appendix 6). But is this a sufficient indication of their production? With a 22 percent larger area of agricultural land than the United States, the Soviet Union produces on the average 50 percent less grain. Appendix 6 is based on the level of grain production in 1981; when including rice production the rankings are occasionally reversed, with China fourth for area, in second place for production behind the United States, and France in the sixth position for production and only fifteenth for area. Indonesia occupies the seventh place in production and the fourteenth in area. This list of the largest producers does not always coincide with that of the largest exporters, but to become an agricultural exporter of international stature, a country must possess extensive agricultural land area and at the same time have a population small enough not to absorb all production. It is useful to think of an agricultural balance sheet. It has not been possible to formulate a real grain balance sheet for the world's principal agricultural countries because it is extremely difficult to obtain a complete picture of the variations in stock and consequently of real consumption. It is possible, however, to calculate "apparent consumption," that is, "production + imports − exports" and to formulate production and consumption increase hypotheses. World import and export capacities can thus be obtained.

The results are eloquent: a country can easily become a large exporter or a large importer of grain depending on production and consumption policies. Assuming that consumption increases by 20 percent over ten years (1.8 percent per year), two separate calculations were performed, one based on stagnant production (0 percent per year) and one based on a growth rate identical to that for consumption (1.8 percent per year). This commonsense analysis is intended solely to detect large potential surpluses and deficits over a period of ten years. The calculation does not pretend to be of scientific value or to function as a means of projection, but its results are striking. It shows the potential sources of new imbalances in 1995. A failure in Chinese agrarian policy would create an additional deficit of 48 million metric tons within ten years; a halt in the growth of European production would produce a deficit only half as large: 24 million metric tons. The Soviet Union could be forced to import an additional 38 million metric tons, but Eastern Europe and the Middle East would import only 16 million and 11 million metric tons respectively; India would import 27 million metric tons, but Japan only 4 million. Turkey and Pakistan, less significant importers, would import 5 million metric tons each.

A halt in the growth of U.S. agriculture would deprive the world of more than 50 million metric tons, but the same situation in Canada would mean a loss of only 8 million metric tons. For Argentina, Australia, and Thailand it would mean a loss of only 3 or 4 million metric tons each.

To complete this analysis of the most important factors in food and agricultural commerce, it seemed appropriate to examine whether smaller countries should be taken into account. Appendix 7 lists some thirty of the importing and exporting countries with more than $1.5 billion in sales or purchases in 1982. Fewer countries have surpluses exceeding $1.5 billion. Among the fifteen countries in this category with surpluses, the following should be noted: Thailand (rice, corn, and cassava); Malaysia (rubber and palm oil); Colombia (coffee and cocoa); Cuba (sugar); and Turkey (cotton, grain, and citrus fruit). Seventeen countries had deficits in excess of $1.5 billion in 1982. The presence on this list of Egypt, Iran, Iraq, East Germany, China, Saudi Arabia, Hong Kong, Switzerland, South Korea, Algeria, Nigeria, and Venezuela indicates the rapid urbanization of formerly agricultural countries and the resulting imbalance in agrarian policies.

This divergent situation in the world's large population zones is orienting commercial currents toward the two traditional importing poles: Europe and Japan, to which the Soviet Union, China, and the OPEC countries must be added more or less permanently. Figure 3.1 provides a simplified picture of this trend.

Trade routes were quite different a century ago. In 1880, the Ukraine was the breadbasket of Europe, and Europe was the principal market. Great Britain in particular drained grain production from the United States, Australia, and Argentina.[14]

The Soviet Union presents a particular problem. Since 1970 its five-year development programs have planned considerable investments in the food and agricultural sector. At the same time Soviet agricultural production goals have been set at modest levels, which, nonetheless, have been met only in exceptional years. Paradoxically, the production units in which investments have been the highest, the sovkhozes and the new "combinats," have had the most difficulty in meeting their goals in livestock production and processing. Meanwhile, 30 percent of overall Soviet agricultural production comes from 3 or 4 percent of the country's total land area, the famous "individual plots." These are family gardens allocated to kolkhozien families, where within an area of one acre an extraordinary system of indoor agriculture has developed. In addition to the cultivation of vegetables and fruit, the system includes intensive husbandry of dairy cattle, hogs, and poultry, based on the use of grain grown on the collective land of kolkhozes. This situation has led the Soviet Union to import each year, in addition to hard wheat to improve the quality of its bread, increasing quantities of feed grain and soybeans to feed an ever-increasing livestock population[15] (see Appendix 19).

On the export side, newcomers to the world agricultural market are relatively few. Brazil is the most noteworthy in the area of vegetable protein, in which Argentina also has recently become a power, notwithstanding its long-held position as a grain exporter. Between them, these two countries furnish 30 percent of world exports of soybeans, leaving the remaining 70 percent to the United States (see Chapter 6, Table 6.6). The EEC, and particularly France, would like to establish a permanent export policy in sugar, wheat, and dairy products toward the Mediterranean zone, which provides Western Europe with petroleum, and toward Eastern Europe. It should be recalled, however, that the EEC is the world's largest agricultural importing zone and that West Germany is the world's largest importer, ahead of the Soviet Union, the United States, Japan, and Great Britain.

Since late 1976, exporters of coffee and cocoa have become important in world agricultural trade. A quintupling of coffee prices and a quadrupling of cocoa prices have shown at this time the strategic importance of these products, now indispensable in the lifestyles of the industrialized countries. The importance in this market of Brazil, the world's largest exporter of coffee, as well as of the Ivory Coast, the world's largest exporter of cocoa, has been reaffirmed.

With respect to sugar, Cuba's position has become less negligible since the embargo on Cuban sugar has been progressively lifted. The same is true of South Africa and Zimbabwe. The European Community is both an exporter of beet sugar and an importer of 1.4 million metric tons of cane sugar from ACP countries (Africa, the Caribbean, and the Pacific, including Mauritius and Jamaica) with which the Community is linked contractually by the Lomé Convention. But as emphasized by Theodore Schultz, Nobel laureat in economics in 1979, four-fifths of the world's production of sugar is improperly located.

Malaysia has become the world's largest exporter of vegetable oil in the last few years owing to its production of palm oil (2.7 million metric tons in 1982), and Brazil occupies third place with 800,000 metric tons owing to its production of soy oil.

Finally, India should be mentioned because since 1975 it has succeeded in stocking, and at times even exporting, grain, sugar, and peanuts. This repre-

FIGURE 3.1 Food and agricultural trade routes in 1977 and 1982 (figures within parentheses are for 1977)

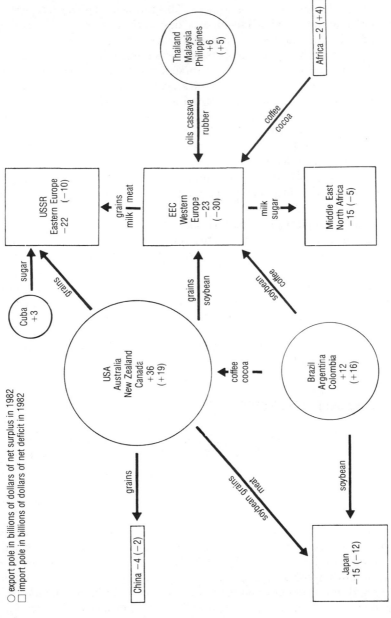

○ export pole in billions of dollars of net surplus in 1982
□ import pole in billions of dollars of net deficit in 1982

Source: FAO, *Trade Yearbook,* 1977 and 1982.

57

sents considerable success given the effect droughts in this country have had on international trade and food assistance. This foresighted policy has permitted India to resort only modestly to imports, despite the extremely serious drought that afflicted this country in August of 1979.[16]

Appendix 8, however, demonstrates the weight of the United States, which with 18.1 percent of world exports estimated in 1982 at $211 billion is the only country to possess a share of agricultural world trade in excess of 10 percent. The exceptional position of the United States is striking. Among the seven principal food product groups traded on world markets, the United States controls 40.2 percent of world commerce in grain, 62.1 percent in oilseeds, 12 percent in oil (behind Malaysia), and 7.2 percent in meat (behind the Netherlands, Australia, and France). All U.S. competitors lag far behind: Canada and France each accounts for about one-fourth of the share held by the United States in world grain exports, and Brazil accounts for only one-fourth in oilseeds.

Only in tropical products such as coffee, cocoa, and tea, and in semitropical products such as sugar, is the United States absent from world markets. Chapter 4 will discuss how the United States failed to capture a share in the dairy market.

In view of these figures, one might speculate that the U.S. agricultural sector has actually been strengthened by the challenges mounted by other new entrants, such as Brazil and Argentina in the soybean market, or by other unpredictable and irregular importers, such as the Soviet Union. Each new challenge seems only a pretext to increase the response capability of American agriculture to supply new deficits in foreign markets.

CONDEMNED TO EXPORT

Under this avalanche of evidence, it is difficult to deny the strength of American agriculture. Nonetheless, between World War II and 1972, U.S. world economic presence was defined primarily by its industry, its currency, and occasionally its military. This apparently paradoxical situation is explained by the marginal role played by exports until 1962 in the American agricultural economy.

What would have been the "normal attitude" of this country? The United States accounted for the major portion of world oilseed and grain production. It would have been logical, then, for the United States to profit from this situation by imposing each year a price on world markets which would have preserved this monopoly and maximized the value of its production. If, for example, in a given year an exceptional harvest in Canada or Argentina resulted in increased competition, the United States would have had an economic interest in lowering the world price to defend its position. The only danger would have resulted if a large Canadian harvest occurred in the same year as a particularly bad U.S. harvest. Yet, even in this case, the American producer would have profited by selling at the world price and then simply reaffirming his leading position the following year.

In fact, the existence of a world monopoly in favor of the American producer did not begin until the 1960s. Of course, exports were already an important component of agricultural policy, particularly after the passage of the famous Public Law 480 in 1954. But this law played only a quantitative role: foreign sales were a privileged outlet for the surpluses accumulated by the American government. One of the apparent paradoxes in U.S. agricultural policy, then, was that for the twenty-five years that marked the greatest development of production in its history, the United States sought only to protect its domestic market rather than to shape world commerce to its advantage. Only since the 1960s have agricultural exports become a vital necessity for the United States.

A Delayed Awareness of Agricultural Export Possibilities

Between 1948 and 1954, agricultural exports played a very minor role in the United States. In 1954, their importance appeared to increase, but in reality it was simply the result of an extensive government policy.

Figure 3.2 provides a graphic illustration of this phenomenon. Its construction calls for a number of comments. The overall balance of trade for American agriculture can be calculated in two ways, according to the definition of imports adopted by the USDA. This agency distinguishes between complementary and supplementary imports. Complementary imports include products that are not produced within the country (rubber, coffee, cocoa, silk, bananas, tea). Supplementary imports include all other products that are imported in competition with American agricultural production. The logic behind this distinction is clear. Complementary imports are irreducible, or in other words, supply elasticity for these products is almost zero, no matter what price is charged in the United States. Supplementary imports, on the other hand, can be substituted for local production, provided that prices are sufficiently low to stimulate demand. To summarize, the following trade balances can be defined:

B1: total exports − complementary imports = degree of dependence.

B2: total exports − supplementary imports = competitivity index of American agriculture.

B3: total exports − total imports = contribution of the agricultural sector to the balance of trade.

Figure 3.2 shows balances B2 and B3. B1 may of course be obtained by subtracting B3 from B2.

It must first be observed that until 1956 the total B3 balance was permanently negative. And yet this balance encompassed all exports in the form of food assistance. In reality, it was not until 1960 that this tendency was reversed in a durable manner, resulting in considerable surpluses in recent years. From 1957 to 1960 overall equilibrium was fragile and was maintained only by massive, highly subsidized government exports.

FIGURE 3.2 U.S. agricultural and food trade balances, 1948–1978

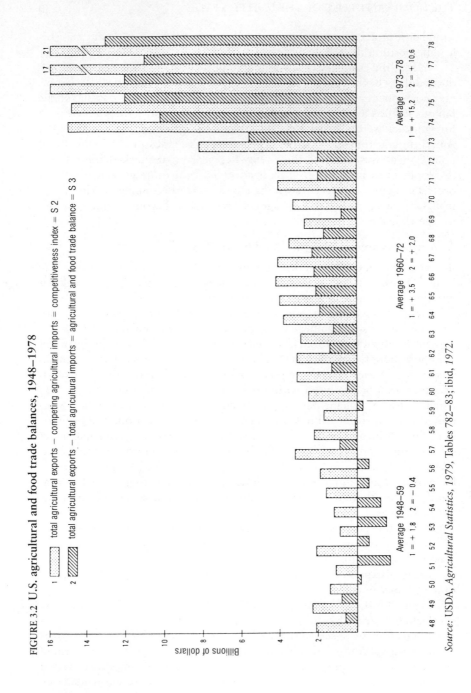

1 ▢ total agricultural exports – competing agricultural imports = competitiveness index = S 2

2 ▨ total agricultural exports – total agricultural imports = agricultural and food trade balance = S 3

Billions of dollars

48 49 50 51 52 53 54 55 56 57 58 59 60 61 62 63 64 65 66 67 68 69 70 71 72 73 74 75 76 77 78

Average 1948–59
1 = + 1.8   2 = – 0.4

Average 1960–72
1 = + 3.5   2 = + 2.0

Average 1973–78
1 = + 15.2   2 = + 10.6

Source: USDA, *Agricultural Statistics, 1979,* Tables 782–83; ibid, *1972.*

TABLE 3.4. Growth comparison in U.S. production and exports

	A: Export Indices (1967 = 100)	B: Farm Output Indices (1967 = 100)	A/B: Export/ Farm Output Indices (1949 = 100)	Average of A/B	Periods
1949	53	74	100		
1950	46	73	88		
1951	54	75	101		
1952	46	78	82	86.6	
1953	41	79	72		
1954	44	79	78		PL 480 passed
1955	50	82	85		
1956	67	82	114		
1957	73	80	127		
1958	63	86	102	112.8	Food aid era
1959	68	88	108		
1960	84	90	130		
1961	84	90	130		
1962	83	91	127	130.3	Support price reform
1963	92	95	134		
1964	104	95	153		
1965	101	98	145		
1966	107	95	156		
1967	100	100	140	146.4	
1968	100	100	137		
1969	94	102	129		
1970	111	101	153		
1971	111	110	141		
1972	129	110	164		Sales to the USSR
1973	156	112	194		Vote on Agricultural
1974	165	106	217		Act of 1973 (target
1975	166	114	191	202.8	price)
1976	174	117	208		
1977	177	119	205		
1978	204	122	230		Agricultural Act of
1979	211	129	225		1977 (storage price)
1980	246	122	278	243.5	
1981	235	134	241		

Source: USDA, Agricultural Statistics, 1972, Tables 633 and 599; USDA, Agricultural Statistics, 1982.

The change in 1960 is illustrated in Table 3.4, which shows how exports progressed in relation to agricultural production. Taking the year 1949 as a point of reference, this index was only 108 in 1959. The following year it rose to 130, where it remained for four years, climbing suddenly to 150 in 1964, where it again stalled for nine years before rising to 200 beginning in 1973 and to about 250 in 1978.

Figure 3.2 and Table 3.4 show that exports did not constitute a major concern for the American government until the mid-1960s. During this period exports were a primary objective only in the sense that it was necessary to find an outlet for stocks accumulated after large harvests, but still not in the economic sense of the term "exports"—that is, foreign sales at the market price in order to earn foreign currency.

First Step toward an Export Policy: PL 480

In 1954, a very important law was passed: Public Law 480. This law played a dual role: it delegated authority for all export operations to a single agency, and it allowed exports to be integrated with internal policy goals. It is not necessary to elaborate on the first point; the preceding tables and figures demonstrate the effect of this consolidation beginning in 1954, and particularly after 1959. PL 480 must consequently be considered as a supplemental instrument in the hands of the secretary of agriculture. The advantage of this instrument is that it is very flexible and that it can be adapted to meet internal objectives as well as the situation in world markets. The mechanism created by PL 480 will be examined in greater detail in Chapter 6, but Figure 3.3 shows that its role in the growth of American agricultural exports became increasingly less significant beginning in 1962. Over a period of eleven years, from 1963 to 1974, the proportion of American exports conducted under PL 480 decreased from 30 percent to about 5 percent, and it has remained constant since 1975. The period between 1960 and 1964, therefore, served as a bridge, allowing food assistance to be gradually replaced as the primary force in export expansion.

The Historic Decision to Lower the Support Prices

It is difficult to say whether government measures in favor of the development of food and agricultural exports preceded the actual growth in such exports. The most important measure was taken by the Kennedy administration in 1962, two years after the appearance of a positive agricultural trade balance. The new measure consisted of lowering the support prices for the principal agricultural products—wheat, corn, and barley—to world price levels so as to permit exports without government subsidies. In exchange for the drop in domestic prices, the American government agreed to provide direct aid to farmers, since the sales price of their production was now subject to fluctuations in the world market.

Ten years later, the Williams and Flanagan reports, as well as the comprehensive new Agricultural Act of 1973, accompanied the growth in exports linked to the opening of the Soviet market.[17] For the first time, an evaluation of

FIGURE 3.3 The importance of food aid in American exports, 1949–1977

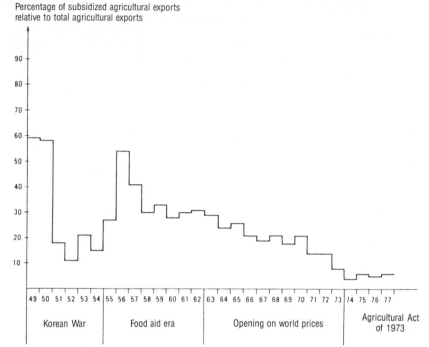

Percentage of subsidized agricultural exports
relative to total agricultural exports

Korean War | Food aid era | Opening on world prices | Agricultural Act of 1973

Source: Willard Cochrane and Mary Ryan, *American Farm Policy, 1948–1973* (Minneapolis: University of Minnesota Press, 1976).

the growth potential of American exports was attempted in these reports. It was conducted, however, somewhat after the fact, since in 1973 and 1974 the value of exports increased by 59 and 65 percent respectively, and the balance increased from +$2 billion in 1972 to +$12 billion in 1974.

In 1979, Secretary of Agriculture Robert Bergland predicted that exports would increase to $32 billion,[18] producing a positive balance of $16 billion. In 1980, export projections established before the Soviet grain embargo were for $38 billion with a positive balance of $20 billion. This projection was met, despite the effects of the embargo; exports reached $40 billion and the balance, $23 billion.

In 1981, a record year, exports rose to $44 billion and the balance to $27 billion. In 1982, the effects of a strong dollar and international debt led to the first decrease in American agricultural exports: $39 billion with a balance of $23 billion. In 1983, the recession continued, reducing exports to $35 billion and the balance to $18 billion. In 1984, there was a recovery, bringing exports back up to $40 billion and the balance to $22 billion.

Not until 1962, therefore, did American agricultural policy place the same priority on nonsubsidized exports as it had on the maintenance of farm in-

TABLE 3.5. Principal clients for U.S. agricultural exports, 1981 and 1974 (in millions of dollars)

Client Countries in Decreasing Order in 1981	1981 Exports (October 1, 1980– September 30, 1981)	1974 Exports (July 1, 1973– June 30, 1974)	Rank in 1974
1. 10-member EEC	8,757	5,400	1st
2. Japan	6,706	3,353	2d
3. Mexico	2,732	610	6th
4. South Korea	2,137	661	4th
5. China (Mainland)	2,118	0	–
6. Canada	2,022	1,195	3d
7. USSR	1,573	509	8th
8. Taiwan	1,105	518	7th
9. Spain	1,054	644	5th
10. Egypt	951	264	12th
11. Venezuela	898	248	14th
12. Brazil	843	369	9th
13. Portugal	764	150	25th
14. Poland	700	306	11th
15. Saudi Arabia	491	92	33d
Total U.S. exports to all countries	43,780	21,293	

Sources: USDA, Agricultural Statistics, 1982, Table 732; USDA, Agricultural Statistics, 1977, Tables 775 and 774.

come. From that year forward, the internal American market has had a symbiotic relationship with the world market: world price fluctuations have an immediate effect on internal prices. This symbiosis became complete in 1972, when the Nixon administration eliminated export subsidies for American wheat, which resulted in the maintenance of $67 per metric ton as the very moderate price of wheat delivered to Rotterdam. The massive purchases by the Soviet Union at this artificial price had the effect of increasing domestic prices paid by American livestock farmers, placing the burden of the increasing cost of export subsidies on the American taxpayer.

Subject to this open-door trade policy since 1962, and even more since 1973, American farmers began producing more and more for export.

Condemned to export in order to survive, American farmers have become dependent not only on national but on international clients as well. But the number of international clients can be counted on one hand, and since the products in question are also very limited in number, this double concentration leads to a virtuous circle: the export of agricultural products and the expansion of exports have become an inescapable imperative for the United States. This phenomenon will be examined in the following section.

TABLE 3.6. U.S. wheat, corn, and soybean exports (millions of dollars)

	1972[a]	1973[a]	1974[a]	1977[b]	1981[b]	1983[b]
Wheat and flour	1,071	2,387	4,737	3,054	8,052	6,166
Corn and coarse grains	1,141	2,346	4,686	5,391	10,393	5,717
Oilseeds and products	2,235	3,507	5,225	6,388	9,305	8,873
1. Total for wheat, corn, and soybeans	4,447	8,240	14,648	14,833	27,750	20,756
2. Total for agricultural and food products	8,046	12,902	21,293	23,974	43,780	39,095
Ratio 1:2 (in %)	55	64	69	62	63	53
3. Total for all products	44,847	57,815	84,924	118,285	229,232	198,350
Ratio 2:3 (in %)	18	22	25	20	19	20

Source: USDA, *Agricultural Statistics, 1977, 1982,* and *1983.*
[a]Year ending June 30.
[b]Year ending September 30.

Clients Too Limited in Number

The top fifteen clients in 1974 and in the record year 1981 are shown in Table 3.5. The top five clients receive more than 50 percent of U.S. agricultural exports; the top ten receive 67 percent, which indicates a significant concentration of exports toward the EEC, Japan, Mexico, South Korea, the two Chinas, Canada, the Soviet Union, Spain, and Egypt.

The situation has evolved extremely rapidly in the last few years and shows the growing importance of the Soviet Union, which occupied third place until the embargo of January 1980, and Iran, which until 1978 was among the top fifteen clients. It also indicates the emergence of new consumers such as Egypt, Poland, Portugal, Mainland China, and Saudi Arabia and the reduction in exports to such traditional U.S. clients as Israel, India, Greece, Pakistan, and Bangladesh.

A Small Number of Key Products

The concentration of exported products is itself great, in that wheat, corn, and soybeans together represent between one-half and two-thirds of food and agricultural exports, whereas since 1973, animal products represent less than 10 percent of the total and finished products less than 5 percent (see Table 3.6).

This concentration on three products is the strength of the American system, which, as will be demonstrated in Chapter 6, is being seen increasingly as the most practical model and easiest to duplicate for the development of intensive poultry, cattle, and hog production.

There is a risk, however, that world agricultural development will be concentrated on technologies that are inadequately diversified. This represents a real danger, which can be compared to a phenomenon observed in genetics when the progress of animal or vegetable varieties leads to the use of an excessively limited number of pedigree lines. It therefore appears advisable for the world's agricultural sector to continue experimenting with alternative models, even if they are less efficient in the short term than the American model.[19]

Agricultural Exports: A Vital Necessity for the United States

Coupled with the development of exports has been an effort to diversify geographically (the Soviet Union, China, the Middle East), yet market penetration has been limited to three products. As a result, more and more acres have been sown for the export market. In twenty years, the proportion of land used for export production has climbed from 20 percent in 1959 to more than 30 percent today. One acre in three, therefore, is now cultivated for foreign consumers. Indeed, in twenty years export markets have grown by 150 percent, but the domestic market increased by only 25 percent.[20]

Agricultural exports have thus become an essential element both for farmers, who derive a large portion of their sales from exports, and for the American economy, which otherwise would experience a much greater trade deficit: 18 to 25 percent of total U.S. exports are food and agricultural products.

Have foreign markets become a necessary production outlet for the United States? Figure 3.4 indicates that agriculture is the uncontested leader in labor productivity gains among all American economic sectors. This is true for each of the periods represented. When this remarkable productivity performance is seen in combination with the depreciation of the dollar, it is easy to explain the formidable leap made by this country in agricultural exports since the early 1970s.

The Sharp Increase in the Dollar Is Endangering American Agricultural Exports

An appreciation in the value of the dollar beginning in 1981, however, has provoked a reduction in food and agricultural exports, posing a dilemma for farmers and the American government. Should the agricultural policy established during the 1960s be maintained, that is, should domestic prices be linked to world prices? Or should protectionist measures and export subsidies be introduced to compensate for the handicap of a strong dollar? The reasons the United States expanded agricultural exports beginning in 1960 still appear to be valid today, even if American monetary policy is placing agricultural policy in serious difficulty (see Figure 3.5 for the inverse correlation between the American agricultural trade balance and the value of the dollar in terms of Special Drawing Rights).

The cumulative effect of developing country indebtness, dollar apprecia-

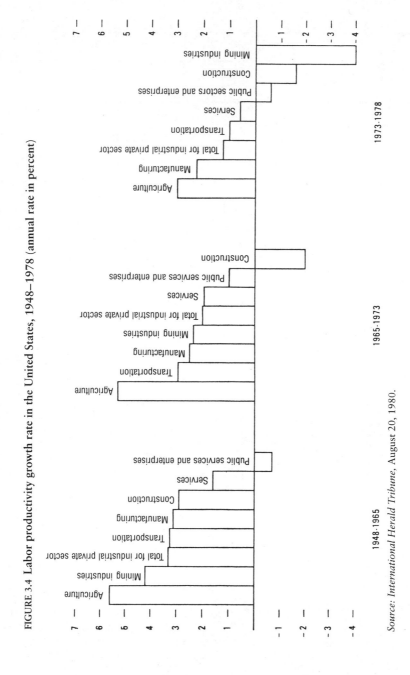

FIGURE 3.4 Labor productivity growth rate in the United States, 1948–1978 (annual rate in percent)

Source: *International Herald Tribune*, August 20, 1980.

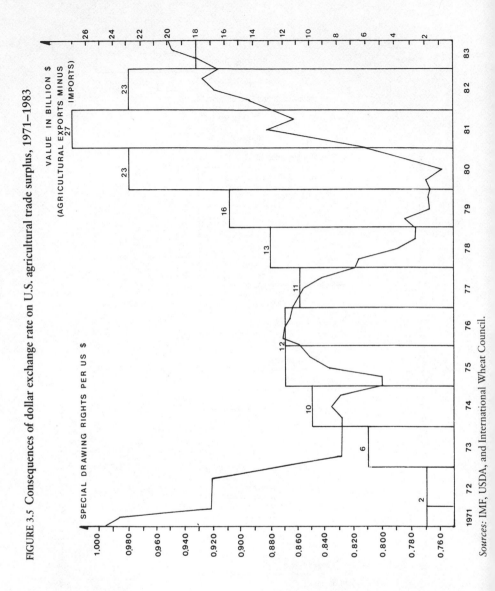

FIGURE 3.5 Consequences of dollar exchange rate on U.S. agricultural trade surplus, 1971–1983

Sources: IMF, USDA, and International Wheat Council.

tion, and increase of support prices has resulted in the United States having a lower share of the world market for agricultural products: 25 percent in 1975, 19 percent in 1980, and 15 percent in 1983.[21]

Three Reasons to Export

An examination of the following studies shows forcefully that exports are no longer simply a matter of choice but of necessity to maintain both a dominant position in external markets and internal equilibrium.

Short- and medium-term projections fully confirm the central role of agricultural exports and particularly of grains and oilseeds. Two studies by the USDA show the extent to which a collapse in American grain exports would be costly for both farmers and taxpayers.

The GOL model, of which two of the principal elements were presented in Chapter 1, has been complemented by a study in 1977 examining the evolution from 1976 to 1981 and 1982.[22] This new illustration of the need for a strong American export policy is based on three hypotheses. Hypothesis 1 is considered as moderate, and the other two represent extreme situations.

Hypothesis 2 is based on both a slow rate of productivity growth (1.3 percent in wheat and 1.8 percent in feed grain) and strong exports (35 Mt of wheat and 48 Mt of feed grain). It is noteworthy that these strong exports have already occurred, for wheat in 1980 and feed grains in 1978. In 1983, despite a decline from 1982, exports of wheat increased to 37 Mt (45 Mt in 1982), and those of feed grain rose to 53 Mt (57 Mt in 1982).

Hypothesis 3 postulates strong growth in yields (2.0 percent and 2.5 percent respectively) associated with lower exports (31 and 42 Mt respectively).

These figures bear out the absolute necessity for the United States to find market outlets and to develop them, for three reasons:

1. A low export level (hypothesis 3) results in a price reduction of 29 percent for wheat and 43 percent for corn as compared with hypothesis 2.

2. Wheat stocks under hypothesis 3 would be twice as large as under the free trade hypothesis and feed grain stocks would be four to six times greater.

3. Budgetary transfers to American farmers would be four times as great solely for income support expenses; isolationism would thus be extremely costly from a tax standpoint.

These short-term projections persuasively suggest to the American government and political circles, whatever their partisan inclination, that a free-exchange agricultural policy is absolutely necessary.

Within the framework of food and agricultural trade dispute, the positions held by farmers and the government of the United States are therefore clear: dominant, imposing, sometimes threatening. Their close symbiosis with foreign markets, although recent, is now solidly established, and it is difficult to imagine that future events could radically alter this situation.[23]

Still, the importance of exports must be viewed in perspective. Even for a large exporter such as the United States, domestic markets account for two-thirds of feed-grain crops. The U.S. agricultural sector owes its extraordinary flexibility and adaptability to this double integration in world and domestic markets.

Part II of this book will show the importance of this double integration and the manner in which U.S. agricultural policy has been progressively modified to reflect America's assumption and consolidation of world agricultural leadership.

Part Two

U.S. AGRICULTURAL POLICY AND STRATEGY

4

The Objectives and Instruments
of American Agricultural Policy

The United States, the home of free enterprise, is also the country in which agricultural policy has been the most extensively studied, applied, and reformed. One of the sectors in which the dynamism of the American economy has been most manifest, agriculture, has also benefited from constant government efforts to balance agricultural production with domestic and international demand.

The purpose of this chapter is to follow the development of an increasingly efficient agricultural policy, which has allowed American farmers to contend with an increasingly unpredictable and diversified domestic and international environment. An examination of the institutional aspects of this policy that allows maximum flexibility to American agriculture will provide a view of the forces involved, their interrelationship, and their capacity periodically to bring about legislative reform.

An economic analysis of the instruments of American agricultural policy created by Congress and applied by the executive branch shows how financial and social constraints were foreseen and an evolutionary, incentive-based system was developed. The unusual coexistence of protectionism, particularly with respect to dairy products, within an apparently free trade framework, poses a problem of international credibility.

THE INSTITUTIONAL FRAMEWORK OF AMERICAN
AGRICULTURAL POLICY

Both the complexity and the extraordinary capacity for change and reform of the American agricultural policy are astonishing. How was it possible to negotiate national consensus allowing transition from the laissez-faire policies of 1910–20 to the Roosevelt policies of 1933–42, which instituted a system of strict controls on cropland area? How were policy makers able to obtain subsidies necessary for the free transfer of 4 to 6 million metric tons of grain each year to the developing countries? How were they able to persuade farmers to accept a reduction by one-half in government-supported farm prices be-

TABLE 4.1. Legislative evolution of American agricultural policy, 1933–1985[a]

Agricultural Acts	Monitoring the Land Area Sown	Support Price	Direct Payments	Food Aid		Presidency	Senate	House
				To the Developing Countries	To the U.S. Poor			
Agricultural Adjustment Act of 1933	Yes	Corn = 60% of the parity price	Yes, for land areas taken out of production	No	Yes	D	D	D
Agricultural Adjustment Act of 1938	Yes, 8 million ha taken off of wheat production, 1938–1943	Corn = 75% of the parity price; cotton and wheat = between 52% and 90% of the parity price	Yes, represented 35% of the sales receipts in 1939, 13% in 1941	No	Yes	D	D	D
War legislation: May 26, 1941	Yes	Max. 85% of the parity price	No	No	Yes, 4 million persons and	D	D	D
October 2, 1942	Yes	Min. 90% of the parity price	No	No		D	D	D
February, 1943	No	Min. 90% to 92.5% of the parity price	No	No	66,000 school-children	D	D	D
Agricultural Act of 1948	No	Max. 90% of the parity price on a decreasing sliding scale	No	Yes (Marshall Plan)	No	D	D	D
Agricultural Act of 1949	Yes (but not applied)	75 to 90% of the parity price (actually 90%)	Yes	Yes (Marshall Plan)	No	D	D	D

(continued)

War legislation (Korea) July 17, 1952	Yes	Yes	90% of the parity price	No	No	D	R	R
Agricultural Act of 1954	Yes	Yes	75–90% of the parity price	No	No	R	D	D
Public Law 480 of 1954				Yes (30–50% of exports)		R	D	D
Agricultural Act of 1956	Soil Bank (11 million ha)	Yes	75–90% of the parity price	Yes	No	R	D	D
Agricultural Act of August 1958	Soil Bank	Yes	Corn: min. 65% of the parity price or 90% of the average price of 1957–56–55	Yes	Yes	R	D	D
Executive Order of January 20, 1961					Yes	D	D	
Feed Grain Act of March 1961	Yes, 20% in reserve (10 million ha)	No	Corn: 74% of the parity price	Yes	Yes	D	D	D
Agricultural Act of 1962	Yes, and removal of the national min. land area for wheat	No	65–90% of the parity price	Yes	Yes	D	D	D
Cotton-Wheat Act of 1964	Yes	Yes, certificates of direct aid are given to pay the parity price for	The national share is paid at the parity price; the surplus share	Yes	Yes	D	D	D

TABLE 4.1 (*Continued*)

Agricultural Acts	Monitoring the Land Area Sown	Support Price	Direct Payments	Food Aid To the Developing Countries	Food Aid To the U.S. Poor	Presidency	Senate	House
		at the world price	the portion consumed nationally					
Food and Agriculture Act of 1965	Yes, "Cropland adjustment" (14 million ha)	Yes	Yes	Yes	Yes	D	D	D
Agricultural Act of 1970	Yes, annual "set-aside" (24 million ha in 1972)	Yes, wheat = $1.25/bu, Corn: $1.00/bu min.: max.: 90% of the parity price	Yes, but limited to $55,000 per farm and per product. The payments are the difference between the parity price and the market price, applied on the national share of the production	Yes	Yes	R	D	D
Agriculture and Consumer Protection Act of 1973	Yes, but not applied	Yes, wheat: min: $1.37/bu max: 100% of the parity price Corn: min. $1.10 max. 90% of the parity price	Yes: deficiency payments from the "target price" and for the overall reference production and no longer from the parity price;	Yes	Yes	R	D	D

76

Act						
Agriculture and Consumer Protection Act of 1977	Yes	Cotton: 90% of the average price of the last three years	target prices: wheat: $2.05/bu corn: $1.38/bu cotton: $0.38/lb; payments limited to $20,000 per farm	Yes (17 million people involved)	D	D
Agriculture and Food Act of 1981		Yes + premium and a medium-term stocking loan (3 to 5 years) (farmers-owned reserve)	Yes, but limited to $20,000 per farm in 1977, $40,000 in 1978, $45,000 in 1979, and $50,000 in 1980 and 1981		R	D
Executive Order of January 11, 1983	Yes + payment-in-kind (33 million ha)	Yes	Yes, but "payments-in-kind" not submitted to the $50,000 per farm limit		R	D
Emergency Act of April 6, 1984			Target prices are frozen to minimum 1983 levels		R	D
Food Security Act of 1985	Yes, with minimum acreage reduction when stocks are high	Yes, loan rates are set at 75–85% of average market prices	Yes, but target prices are reduced 10% and $50,000 ceiling is softened	Yes	R	D

ᵃAbbreviations: D = Democrat, R = Republican, bu = bushel, lb = pound, ha = hectare.
Source: USDA, ASCS, *Compilation of Statistics, Agriculture Handbook No. 746,* rev. 1978.

tween 1962 and 1964? How was it possible to decontrol the planting of grain and oilseeds in 1970 and 1973 after years of production quotas and to create a system of farmers-owned stocks in 1977? And above all, how was it possible to persuade the variety of interest groups (farmers, dealers, processors, government organizations, and others) that these variations constituted optimal adjustments to new market situations rather than an inability to settle once and for all on a consistent policy?

There are no simple answers to these questions. There are, however, three elements that characterize the complex interaction between the interest groups that participate in the elaboration of American agricultural policy: (1) the regularity of revisions in agricultural legislation, which occur nearly every four years; (2) the major role in defining alternative policies played by university rural and agricultural economics departments in all fifty states; and (3) the systematic organization of hearings to achieve a consensus among the various interest groups.

Table 4.1 shows the succession of farm bills. A remarkable regularity is evident: each law is replaced almost every three, four, or five years by new legislation, which generally is presented to Congress by a new president one year after his election. Newly appointed agricultural authorities are thus given the opportunity to institute the reforms they have designed during their period with the opposition party. The Agricultural Adjustment Act of 1933 (following the election of Franklin Roosevelt), Public Law 480 (Eisenhower), the Agricultural Act of 1962 (Kennedy), the Food and Agriculture Act of 1965 (Johnson), the Agricultural Act of 1970 (Nixon), and the Agriculture and Consumer Protection Act of 1977 (Carter) were all elaborated in this fashion. The Agriculture and Consumer Act of 1973, however, which profoundly modified previous law, constitutes a major exception to this pattern. This reform was introduced not by the Nixon administration, which at the time was paralyzed by Watergate, but by the Senate Agriculture Committee, which had forged a cooperative relationship with Secretary of Agriculture Earl Butz. The result was adoption of the "target price," indexed on the cost of production, as a replacement for the "parity price," based on purchasing power in 1910–14. This example illustrates a phenomenon clearly shown in Table 4.1: the House of Representatives and the Senate have been dominated almost without interruption by the Democratic party since 1932. Any comprehensive piece of agricultural legislation is the result of bipartisan compromise, and it is not uncommon for a Democratic president to have greater difficulty in obtaining congressional cooperation, even when Congress is dominated by his own party, than his Republican successor, who often enjoys the support of an agricultural constituency weary of government intervention. Another exception concerned the PIK (payment in kind) program applied for the 1983 harvest. The rules were issued by President Ronald Reagan without any need for a law to be passed by the House of Representatives, which was dominated by Democrats.

The considerable role played by the political process in the design of government intervention in agriculture is certainly one of the most significant characteristics of U.S. agricultural policy. A book by R. Talbot and D. Hadwiger provides a remarkable examination of the details of this process. To understand the often violent reversals in fundamental policy options, it is useful to examine the way in which a majority is obtained. Unanimity in favor of price supports, embracing both ends of the political spectrum, is achieved rapidly. In 1948, "the question whether government should or should not support the market was already settled, at least among farmers and their unions. Even organizations which profess a veneration for the free market, such as the American National Cattlemen's Association, do not consider it a matter of absolute dogma. They desire a free market for livestock but not for livestock feed. A true free market would bankrupt the American agricultural system given current world market conditions."[1]

Unanimity has not been achieved, however, on the best means of supporting prices. The history of American agricultural policy since 1948 is one of the most fascinating subjects an economist can study. It represents a concrete illustration of a debate central to theoretical discussions since the beginning of the twentieth century: What is the most efficient method of government intervention to correct the adverse effects of a free market? Policies targeting prices (subsidies to finance stocks, interest rate rebates, and so on)? Policies targeting income (direct aid, deficiency payments, and so on)? Or policies affecting production quantities (input or output quotas, set-aside premiums)?[2] The political aspects of this question will now be examined, followed by an overview of the economic aspects.

After World War II, party lines were relatively clear: Republicans argued in favor of a free market and Democrats in favor of government intervention. The Republican position mirrored that of the American National Cattlemen's Association and the Farm Bureau Federation. The Democratic position favored price guarantees for producers in the form of government purchases and production quotas. The Brannan Plan, named for the secretary of agriculture in the Truman administration, inspired the market policies applied from 1948 to 1954. It was based on a relatively high support price level, without going so far as to control production. Under Eisenhower, Secretary Ezra Taft Benson introduced a new philosophy of price guarantees, illustrated by a quote taken from Willard Cochrane and Mary E. Ryan: "This philosophy that the government knows better than farmers how to run a farm is more than theoretical."[3] This approach led to a policy applied between 1954 and 1960 which slowly lowered the level of support. Unfortunately, productivity gains by American farmers were so great that government stocks increased rather than decreased.

The same debate, between the same groups, liberals against conservatives, Democrats against Republicans, government interventionists against proponents of an assisted free market, took on considerable importance during the Kennedy-Nixon presidential campaign of 1960. The Orville Freeman Plan, named for President Kennedy's new secretary of agriculture, is a typical approach in which farm income is increased through higher prices in exchange for strict controls on production. Economic theory demonstrates why such a plan is doomed to fail.[4]

Identical divisions exist within the agricultural organizations. The American Farm Bureau claims more than 1.8 million farm and rural family members (rural families subscribe to its mutual insurance system). It is the largest and most dynamic agricultural association. Traditionally conservative politically and economically, this association works closely with the Republican party, favoring less government intervention in agricultural markets. The Grange, with 250,000 rural family members, is the oldest of the agricultural associations, although now in decline. It was organized somewhat like a rural Free-masons association. The National Farmers Union, represented in thirty-five states and well established in the Midwest, includes 250,000 owner-operator families. This association favors greater negotiating power for farmers in their upstream and downstream transactions. The National Farmers Organization, the most recent of the agricultural associations, has made serious inroads in certain depressed areas, such as dairy and poultry farming. This is the "radical" association. It does not hesitate to resort occasionally to direct boycotts or roadblocks in attempts to influence government action and obtain better price supports.

Rural youth organizations also play an important role, but more in the area of education. The 4-H Club is a movement of rural youth consisting of more than 3 million members, principally farm children aged thirteen to seventeen. A highly active organization, 4-H organizes numerous programs to train its members and to conduct projects of collective interest. It also offers many apprenticeships and rural youth exchange possibilities. Future Farmers of America, with 400,000 members aged fifteen to twenty, plays an official role in vocational training of young farmers and is closely linked with teachers of agricultural science.

Individuals have also played a key role, for example, Charles McNary and Gilbert Haugen of the Moline Plow Company in 1924, or Willard Cochrane of the University of Minnesota in 1961, as will be shown further on. It is still more significant that these men later held government posts. More recently, Clifford Harding, secretary of agriculture under President Nixon from 1968 to 1971, his successor, Earl Butz, as well as assistant secretary for Robert Bergland, Dale Hathaway, were all university professors of agricultural economics. This symbiosis contributes to a vigorous public debate, which, in the United States, frequently relates to subtle economic concepts that, in Europe, would be unlikely to stir interest among political circles.

In the final analysis, although Democrats tend to favor more interventionist programs and Republicans a more free market approach, the program ultimately adopted depends primarily on two factors: the personality of the secretary of agriculture and logrolling in the U.S. Congress.

Logrolling, of course, refers to the practice of exchanging votes: "I'll support your bill if you'll support mine." This is a common practice in Congress. Votes become a form of barter, a measure of a legislator's commitment to a given legislative proposal, expressed by the concessions he is willing to make. In addition to logrolling, the academic debate among the various universities during congressional hearings also plays an important role.

Both the administration and the two houses of Congress devote great time and attention to hearings, in which the various interest groups can air their positions. These public forums are held whenever a new law or regulation is being prepared, or when a congressional committee decides to examine a public issue. Most of the participants are either independent public figures or the employees or legal representatives of companies, labor or farmers' unions, or other interest groups. Verbal testimony is generally accompanied by written text, all compiled and published following completion of the hearings.[5]

The lawyers and economists who specialize in representing economic interest groups are called lobbyists because they are often seen in the halls of Congress and the executive departments. Their role, far from dishonorable, is essential to the conduct of hearings, in which they are the "professionals." In short, there is a continual dialogue among participants practiced in the same economic language, ranging from government bureaucrats to congressmen, from private enterprise to labor unions, with lobbyists and congressional staffs playing a central role.[6]

The debate is extremely open, as is amply illustrated in an article by Spitze, referring to the no less than ninety-one interest groups testifying before agricultural price and income hearings held by Congress in February and March 1975.[7] These groups represented extremely exacting and particular interests (such as fertilizer producers' associations, bakeries, and Catholic charitable organizations). On this occasion, logrolling was intense. Representatives with soybean constituencies exchanged votes with representatives from wheat states. In Congress, it is not unusual to see a Republican support a bill sponsored by a southern Democrat in exchange for a vote later on.

The systematic search for consensus is an integral part of the American political system. The process can be lengthy, but the debate is often of the highest quality, resulting in reforms that are broader and bolder than would be considered realistic in Europe.

ECONOMIC ANALYSIS OF AMERICAN AGRICULTURAL POLICY

The first agricultural policy system that comes to mind is the free market, the total absence of government intervention.[8] It has been used little in the United States, even under the most conservative administrations, such as Eisenhower's, or during the period 1973–76. The advantage of this system is its low direct cost to government. This, of course, does not take into account the potential social costs, which can be considerable. The disadvantages are many and easy to illustrate using simple graphs. As shown in Figure 4.1, the combined inelasticity of supply and demand means that a slight change in either causes substantial variations in price. For example, it can be observed that in the United States the gross price index for grain increased from 100 in August

FIGURE 4.1 The free market equilibrium system

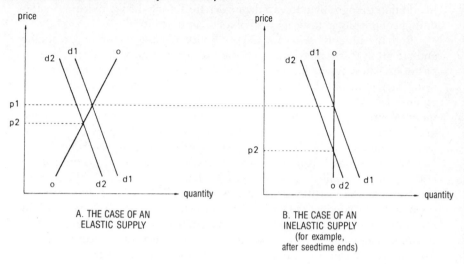

A. THE CASE OF AN
ELASTIC SUPPLY

B. THE CASE OF AN
INELASTIC SUPPLY
(for example,
after seedtime ends)

d1d1 = initial demand curve
d2d2 = weaker final demand curve
00 = supply curve whose slope is equal to
the inverse of the elasticity of the
supply relative to prices

P1 = initial equilibrium price corresponding to
the initial d1d1 demand
P2 = final equilibrium price corresponding to
the final d2d2 demand

1972 to 136 in January 1973 to 291 in October 1974, then fell to 203 in June 1975 (see also Figure 1.1). Furthermore, farmers make decisions on the amount of land to be planted based on price expectations. Because of the seasonal pattern of planting, these expectations produce cyclical movements called "cobwebs" (Figure 4.2). This configuration entails a major disadvantage from the standpoint of supplies: depending on the initially available stocks, on areas planted by producers, and on yields obtained under current climatic conditions, there is no guarantee that supplies during a given year will satisfy local and export demand. This cycle, therefore, is in direct conflict with stable farm income, on the one hand, and guaranteed supplies, on the other. For three reasons (prices which are extremely depressed during periods of overproduction and extraordinarily high during periods of underproduction, cyclical irregularity of production quantities and income, and national security), the social costs of a free market have always been considered much higher than the benefits of a total lack of intervention. This leads to the following observation: a laissez-faire policy with respect to agricultural prices would be justified if resources were allocated optimally in all circumstances. This obviously does not happen.

The question then becomes how to ensure proper remuneration for production inputs. The response is complex because beyond direct support policies it is possible to show that any reduction in uncertainty (through direct

FIGURE 4.2 **Annual adjustment of prices**

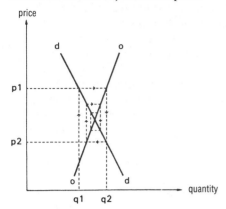

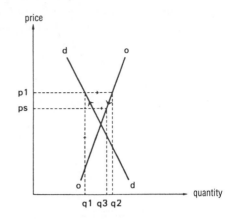

A. ADJUSTMENT SYSTEM IN COBWEBS
(FREE MARKET)

B. PERIODIC ADJUSTMENT SYSTEM
(FLOOR PRICE IMPOSED)

q1 = year's production
P1 = price of the year 1 entailing the production q1
q2 = production of the year 2 planned by the
    farmers keeping the price P1 into account
P2 = price of the year 2 entailing the production q2
Note: Cycle is converging because of high slopes
(i.e., low elasticities) of agricultural supply and
demand curves

q1 = year's production
P1 = price of the year 1 entailing production
    q1
q2 = production of the year 2 planned by the
    farmers keeping account of the price P1
Ps = price of the year 2, the minimum
    imposed by law
q3 = production of the year 3 planned by
    farmers keeping account of the support
    price

price determination, government purchases, or even the use of futures markets) implies a choice in the distribution of production units (and therefore of income). To determine the best policy to guarantee a fair remuneration of inputs, two criteria must be satisfied:

(1) Is the policy efficient in the economic sense (does it minimize costs)?

(2) Is the policy consistent with the society's conception of equity in the distribution of income within the agricultural sector?

It would appear that the only standard that would be both fair and efficient is for agricultural production inputs to receive the same remuneration they would receive if applied to the industrial or service sectors. How to achieve this standard, however, particularly the fairness aspect, remains to be resolved. That producer, farm operator, and farm labor family all often derive their sole income from the farm adds urgency to the question of equity. This point can be easily illustrated by the case in which the government determines the price of a final product above the level required to ensure a balance between supply and demand without intervention. This action has two effects. First, it stabilizes the price. Second, it determines the number of farmers who can remain active because it permits a larger number of individuals to produce at a profit. In this sense, price determination involves an implicit structural element. The level of active government support determines the number of farmers who choose to continue in operation. Government limitations (quotas) on production take this process one step further. By specifying how much each farmer can produce, the government can precisely control the distribution of income within the agricultural sector. These two extremes, free market versus production limitations on individual farmers, have been the subject of long and intense debate, leading gradually to compromises throughout American agricultural history.[9]

The question remains, however, how to ensure proper remuneration of production inputs. The definition of "proper" or "fair" remuneration is itself the subject of extensive debate. Of primary concern is the nature of the concept *parity* or *production costs*. There is also debate over the period of reference, which requires determination of what periods the markets have been in equilibrium.

An initial response to this question was provided in 1924 by the McNary-Haugen Plan, which, although approved by Congress in 1928, was vetoed by President Calvin Coolidge. It provided that the domestic price paid to farmers would be indexed on the price that prevailed in the United States during a reference period when equilibrium was considered to exist. The decade before World War I was chosen, and the buying power of grain that then prevailed is called the "parity price."

The United States has never followed a policy of rigidly setting prices and placing limitations on production volume. Nor did France before its entrance into the European Community. Since the creation of the common agricultural policy in 1962 and up to 1980, however, a rigid price-setting system has been applied to several European products, including grains, dairy products, and beef. Under such a system (see Figure 4.3), any unforeseen increase in supply, caused for example by a greater than average yield, results in a corresponding expansion in public stockpiles because the government-set price cannot be reduced to maintain the initial level.

With such a policy, it is also necessary to ensure that products destined for the world market benefit from export subsidies, called "restitution," which compensates for the difference between the supported price and the equilibrium price in the world market.

The advantages and disadvantages of this rigid system are clear. Prices paid by the consumer and received by the farmer can be stabilized, so farmers can be certain of a given income level, once they know the size of their yield. The system also ensures consistent supply levels. Yet the public cost for such a system is extremely high. If productivity gains are great, the government is rapidly faced with an unsolvable dilemma: supply increases accompanied by stable demand, which implies an ever-increasing public stock. At the initial supported price, the cost of administration is too high and there is a strong temptation to reduce the supported price. In the short term, the effect is contrary to that which is sought: seeing their real income decrease, farmers increase production, although in different proportions. In the long term, they are not willing to accept a lasting reduction in their standard of living. Added to the

FIGURE 4.3 Floor price and the need for public stock

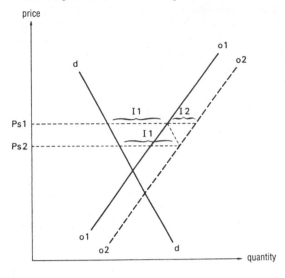

dd = demand curve
o1o1 = initially anticipated supply curve
o2o2 = final supply curve, which was more substantial than projected
P_1^s = initially projected support price
I1 = initially projected public stock
P_2^s = final support price if the government refuses to stock more than I1
I2 = supplementary public stock if the government wants to maintain the initial support price P_1^s
Ps = support price = loan rate

cost of public stocks is the cost of subsidizing exports: upward pressure on the supported price increases the disparity between domestic and world prices. For example, until its abolition in 1972, in the United States the export subsidy for wheat represented up to 65 percent of the domestic price. For that reason, a rigid price system has rarely been applied in the United States, except during periods of war: World War II and the Korean War.

To avoid export subsidies, the McNary-Haugen Plan of 1924 provided that only production for internal consumption would be paid for at the parity price indexed on the period 1905–14. The surplus would be released at the world price, and farmers would receive a blended price. This system was applied for grains before 1962 in France under the name *"quantum:"* beyond a certain quantity produced by each farm, an absorption tax was charged to export at the world price surpluses that were not sold on the domestic market. Export subsidies are then financed by a reduction in the average price paid to producers. The disadvantage of such a policy is that it implies the existence of a public organization constituting a monopoly for collection and often serving as an inefficient substitute for the market. Thus this policy can be applied only in a country where an office such as the Office National Inter Professionnel des Céréales (ONIC) of France or the Canadian Wheat Commission of Canada is the sole transactor with farmers, whom it remunerates at the end of the season according to the prices obtained on the world market.[10]

Before the New Deal, American governments were opposed to any economic intervention. The McNary-Haugen Plan was therefore never applied. But on May 12, 1933, the first agricultural law was passed attempting to overcome the crisis and its drastic effects on agricultural prices, which had collapsed: the Agricultural Adjustment Act of 1933. To restore prices to the "parity" level with the period 1909–14, considered as a period of equilibrium before World War I, a voluntary reduction of cultivated land area was organized by the law based on financial compensation.

A macroeconomic analysis of such a system is fairly clear. As can be seen in Figure 4.4, an increase in the supported price results in a movement along the supply curve, which shifts to the right because of productivity gains. To maintain control over the stock levels (for example, at zero), strict production control is necessary. The higher the level of support, the more restrictive will be the controls. This restrictiveness on production makes analysis of the microeconomic effects of this policy more difficult.

The European Economic Community has been tempted by this policy and has instituted payments for the uprooting of certain fruit trees or vine plants as well as for stopping the supply of milk or the conversion of cows to direct milk feeding of their calves. At one point, under the first version of the Mansholt Plan, the Community considered reforestation of 5 million hectares of cropland.

In the United States, the Agricultural Adjustment Acts of 1933 and 1948 (Hope-Aiken bill), emphasized the double role that should be played by prices: first, as a support instrument to guarantee farm income, and second, as a guide to indicate the quantity of production desired by the government. To this end, the prices of basic products were supported for one year, without variation, at 90 percent of parity. This last sentence encompasses several concepts that will be useful for the remainder of the discussion. The invariable support is explained by the system of crop loans (the nonrecourse loan system). The government announces a price at which it agrees to purchase and stock farm production following harvest. This price is set in dollars per bushel. After harvest, therefore, the farmer has a choice: he can sell his production on the market if the price is higher than that offered by the government, or he can place his production in government stocks and cash in on the price guarantee in the form of a loan at a preferential rate of interest. These stocks are then in the hands of the CCC, the executing agency of the secretary of agriculture. The farmer then has about nine months to decide. At the end of this period, he can leave his production to the government or sell it on the market. This decision is his alone and is of course influenced by the relationship between the market price and the guaranteed price. This system was beneficial for producers of corn, cotton, peanuts, rice, tobacco, and wheat. This list has changed little since 1948. The unit amount of the crop loan (loan rate) thus played the role of a veritable "support price" or floor price.

FIGURE 4.4 Policy for the mandatory reduction of cultivated land areas

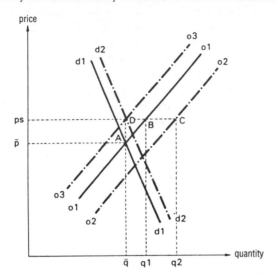

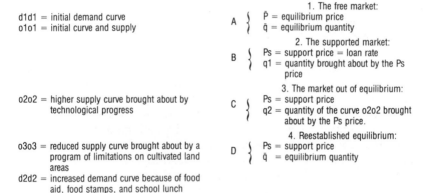

d1d1 = initial demand curve
o1o1 = initial curve and supply

o2o2 = higher supply curve brought about by technological progress

o3o3 = reduced supply curve brought about by a program of limitations on cultivated land areas

d2d2 = increased demand curve because of food aid, food stamps, and school lunch programs

A {
1. The free market:
$\bar{P}$ = equilibrium price
$\bar{q}$ = equilibrium quantity

B {
2. The supported market:
Ps = support price = loan rate
q1 = quantity brought about by the Ps price

C {
3. The market out of equilibrium:
Ps = support price
q2 = quantity of the curve o2o2 brought about by the Ps price.

D {
4. Reestablished equilibrium:
Ps = support price
$\bar{q}$ = equilibrium quantity

The system of direct reduction of cropland was judged unconstitutional by the U.S. Supreme Court, however, and a new policy, now one of the foundations of the American agricultural system, was defined by the Agricultural Adjustment Act of 1938. In place of the rigid parity price system, the 1938 act substituted the concept of "parity income," as the result of an interim law, the Soil Conservation and Domestic Allotment Act of 1936.[11] The level of support (loan) was set in a flexible manner depending on the year between 52 and 90 percent of the parity price (referring to the level indexed on the period 1909–14), in inverse proportion to the foreseeable surplus in supplies. For a normal production year, the price was supported at 70 percent of parity; if production was 75 percent of normal, support increased to 90 percent of parity, and inversely, for overproduction of 30 percent, support decreased to 60 percent.[12] This method is based essentially on the government's use of a graduated scale. It corresponds to a ceiling on quantities stocked, in the absence of any directive other than the "function of government demand."

Indeed, the flexible price system can be interpreted as an effort to increase the elasticity of the demand function confronting producers. Once again, if productivity gains are strong, supplies increase rapidly and, under the system of flexible prices, there is downward pressure on the government price. To compensate their resulting loss of income, the small farmers, with a large share of fixed costs, will increase their production. As a result of this situation, the government cannot refuse to support prices at a level above the one set in the scale. At that point, the advantages and disadvantages are comparable to those in the rigid price system.

Under the effects of World War II, the goal in the United States of restricting production gave way to that of encouraging production, through the use of high guaranteed prices, which persisted until the 1950s.[13] This "war economy" made a return to the rigid price system necessary, with the support price remaining at 90 percent of the parity price for the entire period. Despite a number of attempts, this system remained in place until the end of the Korean War in 1953.

Following the return to overproduction in 1954, a twofold action was undertaken to increase demand and decrease supply. The Agricultural Trade Development and Assistance Act of 1954, better known as Public Law 480, was designed partially to increase demand by creating a vast program of food aid. The Soil Bank, created by the Agricultural Act of 1956, was in turn intended to decrease supply. In 1960, 11 million hectares were voluntarily idled under the terms of five- to ten-year contracts signed with the Soil Bank. Despite these two volunteer-oriented policies designed to shift the supply and demand curves and arrive at a new equilibrium, wheat stocks continued to increase, reaching 38 Mt in 1960, while corn stocks reached 5 Mt in mid-1961.

It appeared necessary to evolve a more restrictive system. The aim of the Kennedy administration beginning in 1960 was to reshape American agricultural policy to absorb the enormous surpluses resulting from the extraordinary gains in productivity after World War II (see Figure 4.5). The first attempt to avoid the shift caused by greater than anticipated yields was to use a system of quotas, linked to a high support price. In its simplest form, this system consists of what are called marketing quotas. This technique involves the setting of production quotas for farmers. Farmers deliver their output, which could be produced only on the land area allocated. In the event that additional land area was planted, penalties were imposed. Such a system is clearly effective in limiting production, but it is very costly to enforce, unless production passes through a set processing point. Even in this case, such quotas are so difficult to divide among producers that their implementation had to be approved by a two-thirds vote by the farmers.

In 1963, when Orville Freeman proposed a plan (the wheat referendum)

FIGURE 4.5 **Wheat stocks from 1947 to 1973**

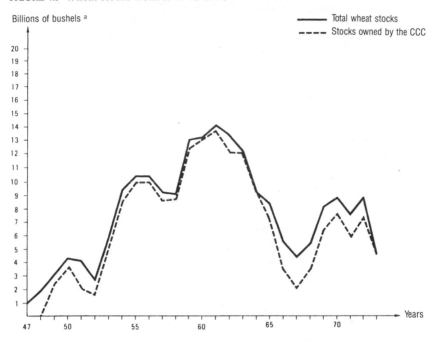

a1 bushel of wheat = 27.2 kg

based on a similar principle, it was rejected by a large majority. As the story goes, Freeman, in his disappointment, fell ill and remained in his office for several days refusing even to consider alternative solutions.

As a result of the farmers' refusal of this plan, the Cotton Wheat Act of 1964 was proposed and adopted. This law almost completely reversed existing trends. The support price (that is, the level of the loan rate) declined to the world level, which in 1964 was about 52 percent of the parity price. To compensate for this decrease in income, farmers who agreed to participate in the *voluntary* program to reduce land under cultivation received a certificate corresponding to the difference in value between the market price and the parity price, but these certificates were awarded only for that portion of the harvest destined for domestic markets, or about 40 to 45 percent of the harvest (Figure 4.6). To finance part of the cost of the direct aid, the processors, or millers, must pay a tax equal to about 30 percent of the parity price. This tax, which is reflected in consumer prices, results in a double price, the domestic price, P_i, which is higher than the support price, P_s, which is close to the world price. The financing of these direct payments is therefore assured half by taxpayers and half by American consumers of flour (see Figure 4.6).

To calculate the direct aid awarded to producers, each farmer was given an

FIGURE 4.6 The 1969 wheat program (support price at the world standard and direct aid for the percentage consumed nationally)

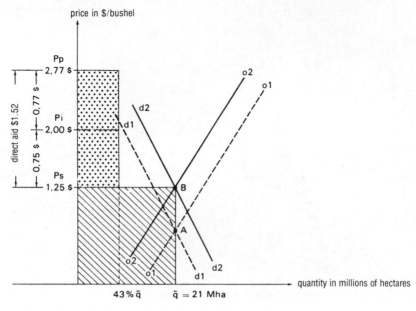

SSSS Receipt of sales taken from the market = Ps × q

░░░░ Direct government aid = (Pp − Ps) × 0.43 q̄
      partially reimbursed by the millers, hence by the consumers

d1d1 = initial demand curve
o1o1 = initial supply curve

d2d2 = demand curve increased under the effect of
       PL 480, school lunch, and food stamp
       programs
o2o2 = reduced supply curve after the voluntary
       set aside of 15 percent of the land areas

A { 1. Free market:
     p̄ = insufficient equilibrium price
     q̄ = equilibrium quantity = 21 Mha

B { 2. Reestablished equilibrium:
     Ps = support price = $1.25/bushel
     q̄ = equilibrium quantity = 21 Mha

Pp = parity price = $2.77/bushel; 43 percent q̄ = quantity consumed nationally
Pi = domestic price = Ps + the millers' certificate = 1.25 + 0.75 = $2.00/bushel

acreage allotment. This allotment was first set at the national level: the land area used to produce the amount necessary for national consumption and "normal" exports. This amount of land was then divided among the states, then among the counties, and finally among individual farmers, taking into account their planting levels measured during a reference period, generally 1959–60. Each farm thus had a quota or "historical base" for each crop, which can be regarded as a "production right." This quota served as a reference for the calculation of the direct aid the farmer received to benefit from the parity price on that part of his historic base which corresponded to domestic consumption.[14]

It required great courage for the American government to shift in four years (1960–64) from a support price equal to 90 percent of the parity price to a level equal to only 52 percent. This policy was so efficient in reducing the burden of financing stocks and replacing that system with one of direct aid to farmers that it was renewed virtually without change in 1965, 1968, and 1970.

The broad consensus which this policy gathered in the United States also resulted from the introduction immediately following the inauguration of President Kennedy of a program for the distribution of food to the economically disfavored. This consumer aid policy, known as the Food Stamp program, will be studied in greater detail in Chapter 5. This program, reapproved with each revision of agricultural legislation, allows representatives of urban districts to rally around legislation favoring farmers.

This system of direct aid to farmers, however, presents a number of disadvantages. The distribution of acreage allotments for each crop resulted in great inflexibility in the choice of annual planting: a farmer with a historical right for wheat could not decide to switch to barley, even if this crop were better suited to his operating system. This is why, during renewal of agricultural legislation in 1970, the Nixon administration introduced a new concept: once a farmer had voluntarily idled (set-aside, or acreage reduction program as it was later named) a portion of his land in compliance with directions from the secretary of agriculture, the rest of his land could be planted as he wished with any grain or oilseed, without forfeiting the direct aid corresponding to his historical right. The author of this concept of set-aside, the economist Grover Chappell, believes it will transform government-supported agriculture into a commercially oriented agriculture that will be much more dynamic.[15] In fact, the assistant secretary of agriculture, Clarence Palmby, explained that he would thereafter establish the support price (the loan rate) just below the world price and no longer just above as had been the case since 1964. This nuance changed the intervention agency, the CCC, from a stockpiling organization into a bank issuing warehouse warrants for crops. Each farmer had a greater incentive to sell at the world price than to place his production in the government stock.

The Kennedy policy established by Willard Cochrane and Orville Freeman and applied since 1964 presented a second disadvantage, which became particularly evident following the price increases caused by the Soviet purchases of 1972: at any market price, as long as it was less than the parity price, farmers would receive direct aid (see Figure 4.6). The scandalous amounts of aid provided in 1972 and 1973, despite unprecedentedly high agricultural prices (which still, however, were lower than the parity price), made change a necessity. The Senate Agriculture Committee, dominated by Democrats, presented to Republican Secretary of Agriculture Earl Butz a new method of calculating direct aid. Butz quickly accepted it because it corresponded exactly to his own concepts. The plan became the Agriculture and Consumer Protection Act of 1973. The reference to parity prices was abandoned, replaced by the target

FIGURE 4.7 Calculation of deficiency payments under the Agriculture and Consumer Protection Act of 1973

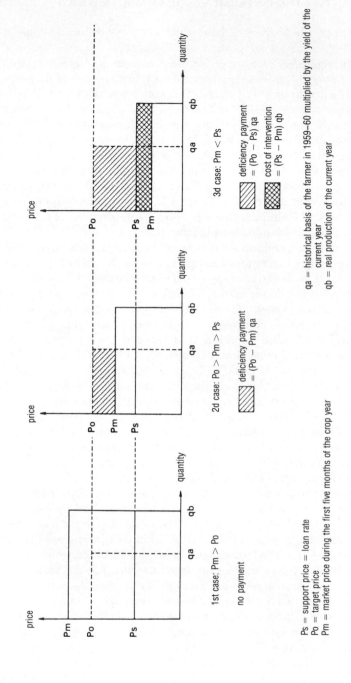

1st case: Pm > Po

no payment

2d case: Po > Pm > Ps

deficiency payment
= (Po − Pm) qa

3d case: Pm < Ps

deficiency payment
= (Po − Ps) qa

cost of intervention
= (Ps − Pm) qb

Ps = support price = loan rate
Po = target price
Pm = market price during the first five months of the crop year

qa = historical basis of the farmer in 1959–60 multiplied by the yield of the
        current year
qb = real production of the current year

price, which was much lower and based on average production cost rather than on purchasing power in 1909–14. Direct aid was replaced by a deficiency payment, which would no longer be made every year. It was granted only if prices during the first five months of the season were less than the target price (see Figure 4.7). Thus, practically no payment was provided to farmers between 1974 and 1977, as shown in Table 4.2, and deficiency payments from 1977 to 1981 were considerably less than direct aid payments in 1972 and 1973.

It is pointless to examine all the details of the explosion in international commerce after 1973, but following are the main points:

1. The price of wheat (U.S. No. 3 Hard Winter, Ordinary, fob Gulf) increased from $60 per metric ton in January 1972 to $214 per metric ton in January 1974.

2. Corn (Yellow No. 2, fob Gulf) increased from $51 per metric ton to $122 during the same period.

3. The stocks of wheat in the exporting countries decreased from 52 million metric tons to 29 million metric tons.

Within this context of international price increases (which reached far higher levels than the loan rate in the United States) the government had to decide on the 1973 farm bill. American agricultural policy makers were persuaded that a lasting balance between agricultural supply and demand had been reached. Thus they believed that from 1973 on it would be unnecessary to intervene on a large scale in the market to guarantee adequate remuneration for farmers.

Expectations with respect to the decrease in stocks were shown to be correct for all grains. But the almost total disengagement of the Commodity Credit Corporation was even more significant. For wheat, for example, the corporation's purchases fell from 38 Mt in 1961 to zero in 1974. The cost of interventions was also considerably reduced because in a single year (July 1, 1973, to June 30, 1974) the value of stocks held by the corporation fell from $213 million to $60 million for feed grains and from $195 million to $26 million for wheat.

To respond to this new situation, new policy instruments were necessary. The U.S. situation can be summarized as follows: (a) this country possesses a comparative advantage in agricultural production; (b) it enjoys a world quasi-monopoly; and (c) it must be able to derive maximum advantage from a liberalization of trade, particularly from the opening of economic borders both for agricultural and industrial products.

American negotiators then adopted the following principles in all internal or international debates on trade liberalization:

1. Greater stability in trade is desired, but regulation should relate to quantities rather than prices.

2. The most important aspect of market management is information (particularly projections); this is associated with the idea that any market must be able to project prices when it possesses perfect information.

3. Trade within markets free of economic barriers allows for coherence in

TABLE 4.2. Government intervention in agriculture: Direct aid and deficiency payments to farmers (CCC price support and related expenditures), 1956–1988 (in billions of 1983 dollars)

Year	Values of Public Stocks[a]	Set-aside Land Areas (M ha)	Total Payments to Farmers[b]	Share of Major Crops[c] (%)	Share of Other (Dairy)[d] (%)	Observations
1956–1960 average	24.7	9.6	5.7	n.a.	n.a.	Soil bank
1961–1965 average	22.4	20.8	7.2	70	30	Direct aid cal-
1966–1970 average	11.9	21.6	8.0	81	19	culated on the parity price
1971	11.3	15.2	6.4	56	44	
1972	7.4	24.8	9.5	83	17	
1973	7.6	8.0	8.0	59	41	
1974	3.2	1.2	2.0	n.a.	n.a.	Deficiency
1975	1.1	0.8	1.1	75	25	payments era
1976	1.2	0.8	1.8	36	64	with $p_m > p_o$
1977	1.8	0.4	6.2	74	26	with $p_m < p_o$
1978	6.1	7.2	8.6	59	41	and with the
1979	7.3	5.2	4.9	46	54	system of
1980	6.1	0	3.3	79	21	"farmer-
1981	8.7	0	4.8	34	66	owned re- serves"
1982	9.2	4.4	12.0	78	22	Farmers-
1983	16.9	30.8	18.8	67	33	owned re- serves and PIK era
1984 (Pre- vision)	n.a.	11.0	6.0	32	68	Freezing the target price
1985–1988 projection	n.a.	n.a.	12.1	n.a.	n.a.	Reduced loan rate

Source: Congressional Budget Office, Policy Options for Contemporary Agriculture (Washington, D.C., 1984).

[a]Commodity price support loans and government-owned inventories at start of fiscal year.

[b]Commodity Credit Corporation price support and related expenditures by fiscal year, in 1983 constant prices (direct aid before 1974, deficiency payment after 1974, disaster programs, paid diversion program).

[c]Major crops: wheat, feed grains, rice, upland cotton, soybeans.

[d]Other: mainly dairy program expenses but also wool (36 M.$ in 1979), potatoes (24 M.$ same year), honey (3 M.$), forests (15 M.$), conservation and erosion fight (245 M.$).

n.a.: not available

p_m: market price

p_o: target price

national goals and stability in the international order, and farmers' income is guaranteed thanks to exports; for consumers and producers alike, prices are an indispensable signal orienting supply and demand; stocks, if managed from a perspective of supplying world markets, minimize costs for each government and each of the various operators: farmers, ranchers, processors, exporters, importers, and consumers.

All of these principles are naturally found in the provisions of the Trade Act of 1974 and the Agriculture and Consumer Protection Act of 1973.

Domestically, the introduction of deficiency payments constituted the greatest innovation in 1973. Through this system, the programs stopped focusing on protecting income levels. The support price (loan rate) was still slightly lower than the world price, but a target price triggered compensation of producers if the price fell. Thus modified, the mechanism of set-aside took place at an extremely low price level. In exchange, an insurance system was developed to attempt to protect farmers against the risks of extreme variations in climate (disaster payments).[16] Farmers are now exposed to the rigors of variations in demand, but in exchange they receive extensive protection against the uncertainties in supply conditions.

Beginning in 1975, tension in world markets began to ease, and in 1976 and 1977 public stocks were still maintained near zero (see Table 4.2). This situation caused concern among farmers and threw doubt on which period should be considered "normal," 1972–74 or 1975–77. This perplexity can also be found in the modifications made by the 1977 farm bill, which reflected the conflicting desires to return to domestic intervention and to continue with free market policies.

The Agriculture and Consumer Protection Act of 1977 retained the spirit of the 1973 law, but under the determination of President Jimmy Carter and the influence of Senator Hubert Humphrey, it introduced a new instrument for quantitative regulation: medium-term stockpiling for periods between three and five years, known as the farmers-owned reserve program. Indeed, the loan rate system allowed American farmers to finance their stocks on the farm at the support price level for a maximum of nine to twelve months, or until the next harvest. In exceptional cases, the government could authorize extension of loan rates for a new period of twelve months. In the new system established in 1977, whenever prices dropped below a "reserve release price," set at a level between 140 and 160 percent of the support price for wheat (125 percent of the support price for corn), farmers could take out a reserve loan at an advantageous level, higher than the regular loan rate, and get a stockpiling premium. Farmers could draw down their stocks as soon as market price levels exceeded the reserve release price. They were obliged to repay the reserve loan once price levels exceeded a recall price set for wheat at 175 percent of the support price (145 percent for corn) and forfeit the stockpiling premium if they did not sell their stocks immediately (see Figure 4.8). This entire mechanism for creating and releasing reserves was conceived to stabilize the market price at a level near the reserve release price, or, what amounts to the same thing, at a level slightly

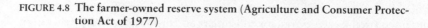
FIGURE 4.8 The farmer-owned reserve system (Agriculture and Consumer Protection Act of 1977)

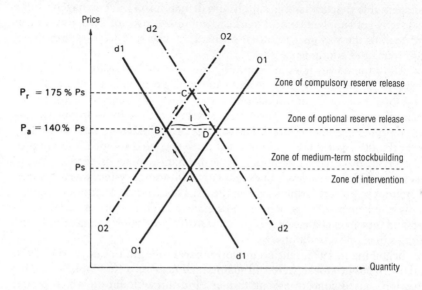

1. Initial market equilibrium:
o1o1 = initial supply curve
d1d1 = initial demand curve

A { Ps = Support price = loan rate
qa = equilibrium production

2. Stockbuilding:
o2o2 = supply curve reduced by building up the medium-term reserve
I = medium-term reserve

B { pa = reserve release price
qb = reduced quantity available for the market

3. World shortage:
d2d2 = higher demand curve caused by an unforeseen world shortage

C { Pr = recall price for the reserves, causing the compulsory release from storage
qc = quantity available for the market at the Pr prices

4. Releasing stock:
o1o1 = supply curve increased by the release from storage of the reserve

D { Pa = reserve release price
qd = increased quantity from the reserve I available for the market

above the target price (which varied from year to year between 130 and 145 percent of the support price for wheat and between 100 and 115 percent for corn).

It is too early to judge the long-term effects of the 1977 act. But for the present, its relative vagueness in areas generally requiring great precision is striking. The legislators' uncertainty can be seen in the extremely wide latitude allowed the administration, with the exception of such specific areas as prohibition of embargoes. The secretary of agriculture is authorized to modify the reserve release price and thus the conditions for the triggering of the reserve system; he may also determine the production quantity and hence the acreage allotments considered necessary for the calculation of deficiency payments. If a farmer elects to sign a voluntary program agreement, his gains in flexibility, under these conditions, are perhaps illusory because the government determines the optimal mix of intervention instruments. Finally, considerable uncertainty also hangs over the costs created by these new provisions. When the world price falls substantially below the target price, it is entirely possible that the total cost of government response will be high. The OECD estimated this cost at $12–15 billion per year for 1980–84, a level that was exceeded in 1983. Furthermore, it is to be expected that an economy in which all producers are supposed to be close to production equilibrium (absence of a surplus) is difficult to manage once this equilibrium is lost. In other words, it is more "costly" to explain to a large farmer that he must operate below his optimal level than to explain to a small farmer that he must stop producing.

This weakness in American agricultural policy was accentuated by the limitation on direct payments required by the 1970 act. Each farmer may receive no more than $55,000 per year in the form of direct aid or deficiency payment for each of his harvests of wheat, feed grains, or cotton. This limit was lowered to $20,000 by the 1973 act. Without this limitation, the law would not have been approved by representatives of urban states or states with small farming operations, which are sensitive to the inequitable influence of income redistribution represented by payments made in favor of farms with incomes greater than $500,000 per year. In 1982, these farmers accounted for 28 per cent of gross farm income in the United States,[17] and their election to sign for government programs to guarantee income and voluntarily reduce cultivated land area could be withdrawn at any time. Programs to limit cropland area, consequently, must concentrate on the smaller farms, which should set aside a much larger share of their land. At the same time, large farmers who refuse to join programs because they no longer see a sufficient financial advantage in doing so would still profit from the restoration of prices resulting from the set-aside of lands by the small farms. This analysis suggests that there is a basic incompatibility between excessively strict limitations on payments to large farmers and efficient control of cultivated land areas.[18] The American government could therefore be led to allow prices to drop below the level of the target price, make deficiency payments without excessively limiting cropland area, and finance large medium-term stocks. This trend seems to have been followed from 1977 to 1982 (see Table 4.2), even though the 1977 act took a step backward and progressively increased the limit on payments per farm from $20,000 in 1977 to $40,000 in 1978 and to $50,000 from 1980 on. At the time of the PIK program in 1983, President Reagan decided that the limit of $50,000 would not apply to the payments under this program. An inquiry by the General Accounting Office showed that thirty-five farmers had received more than

$500,000 each, and seven farmers had received more than $2 million each. In 1984, when PIK was renewed for wheat, payments made under this program were again placed under the $50,000 ceiling.[19]

One might wonder why Congress and the administration chose to adopt such provisions. It should not be overlooked, however, that the meat and livestock feed producers are the privileged sector. It is therefore indispensable to continue encouraging corn producers, so that they will be able to deliver their production at a price that is not too high for beef, pork, and poultry producers. The possible conflicts between internal and external policy goals given a dominant meat sector would thus be resolved, but this threatens to create new uncertainty if the economic and biological cycles of the livestock do not correspond to the regular, annual cycles in crop production.

What can be concluded about the probable directions of U.S. agricultural policy in the years ahead? Following are the few simple facts that are evident in the evolution of U.S. agricultural policy:

1. The American agricultural productive structure can sustain fluctuations in demand without major damage. Supplies, however, must be stabilized by guaranteeing the availability of inputs at a competitive price.

2. By 1990 or 2000 the world food deficit will be most striking in meat. This is the only product for which regular increases in relative price can be expected.

3. It is crucial for the United States to maximize the value of its agricultural exports to compensate for its deficit in energy.

4. This maximum can be attained only if trade is freed of all institutional barriers. Free trade will allow for vigorous growth in the developing countries and will guarantee that the world price is that charged by producers with a comparative advantage.

The strategy of the Democratic administration, beginning in 1976, took all these constraints into account. It consisted first of encouraging producers of wheat, corn, and soybeans through the establishment of a reserve stock. This policy, which departed from the free market principles previously espoused, could prove to be very costly. But the benefit of providing a less unstable price for livestock producers may justify this cost for a time. Finally, this effort to exploit a comparative advantage is not unilateral: the United States realizes that purchasers of its meat will also want to sell their own products. It is not surprising, therefore, to see the American negotiator, whether he be Democrat or Republican, apply strong pressure to eliminate all customs barriers.

The election of President Reagan in 1980 coincided with a contraction in international markets for American agriculture, provoked by the second oil crisis and the widespread debt of developing and Eastern countries. This market contraction was accompanied by a loss in the competitiveness of American prices because of a substantial appreciation of the dollar (see Chapter 3, Figure 3.5). The result was a decrease in world prices for grain and soybeans and a consequent decrease in the net income of American farmers, which dropped from $32 billion in 1979, a level that had already been attained in 1972 in constant dollars to $22 billion in 1982, despite a considerable increase in income support expenditures in the form of deficiency payments set up by the 1973 act and renewed by the 1977 and 1981 acts (see Figure 4.9). Furthermore, financing for the medium-term stocks, set up by the 1977 act (farmers-owned reserve) and renewed in 1981, despite certain minor modifications desired by the Reagan administration, led to the accumulation of very large stocks both in wheat and corn (140 million metric tons of grain, or a year's worth of exports in 1982), because the participation of American farmers in the set-aside program was very low.

Even though the Republican administration expressed a philosophy of leaving farmers to face the market, it decided on January 11, 1983, to make a bold move with the introduction of the PIK (payment in kind) program.[20] Under this program, Secretary of Agriculture John Block obtained the voluntary set-aside of 33 million hectares of grain land, or 36 percent of such lands in the United States. This represents an area of land larger than the grain lands of the ten members of the EEC.

The mechanism adopted is particularly apt because it consists of indemnifying American farmers who agree to limit the area of their land under cultivation, not in exchange for guarantees of deficiency payments or premiums for nonproduction (the paid diversion program), but by delivering a percentage (from 80 to 95 percent) of the production they would have had on the idled land. At the same time, this payment in kind allows medium-term stocks financed with government assistance to decrease. The results have been remarkably effective: 14 percent reduction in wheat production and 27 percent reduction for corn, cotton, and rice. Wheat stocks diminished by 10 percent, rice stocks by 50 percent, corn stocks by 45 percent, and cotton stocks by 30 percent. Moreover, a reduction of 10 percent has been achieved for land area devoted to soybeans, even though this product did not come under the PIK program. The production of soybeans as a secondary crop, complementing the wheat and corn production which was eliminated through the PIK program, was abandoned. Furthermore, many farmers preferred to return to the production of corn, the price of which was restored by the PIK program, rather than planting soybeans. Soybean stocks thus decreased by 25 percent.

Although the PIK program was effective, its budgetary cost was considerable. Nearly 15 million metric tons of wheat and 45 million metric tons of corn have been distributed to farmers participating in the program. Of course, the program limited the government's cost of financing medium-term stocks in 1984 and 1985, but in 1983 the cost of transferring 60 million metric tons of grain from public to private hands reached $10 billion, which added to the $19 billion engaged for deficiency payments.

Thus the year 1983 will probably go down as the culminating point in agricultural support expenditures in the United States. On April 4, 1984, a few days after the European Community decided, for the first time in its agricultural history, to lower the current value of support prices for European farmers, the U.S. Congress passed a law freezing the target prices for 1984 and

1985 at the minimum level provided for 1983. Because of the greatly diminished generosity of the administrative measures adopted simultaneously for 1984 (their cost would decrease from $12 billion to about $7 billion), only 11 million hectares were voluntarily set aside in 1984 as compared to 33 million in 1983. The 1984 harvest set a new record.

As discussions began on the 1985 farm bill, the United States was therefore confronted once again with the problem of surpluses, as it was in 1954 and 1962. Resorting to ad hoc measures decided from year to year is dangerous because it prevents farmers, the main economic actors, from pursuing a coherent medium-term strategy.

Since the current programs have become less and less capable of maintaining farm income, and excessively high support prices cause a decrease in exports, some reflection has been given to two ideas: (1) farm income insurance, against both low production and low prices, and (2) credit better adapted to promoting exports. Since American and world prices are no longer supported by the American government, this policy would make American exports more competitive than those of their more interventionist competitors.[21]

Little probably will be changed in the now very complete panoply of American agricultural policy tools, and once again, greater latitude will be allowed to the USDA to set the various parameters.

As in 1960, when the newly elected Kennedy administration was faced with the difficult task of lowering price support levels, in 1985 the newly reelected Reagan administration was in the same situation.

The jump in support prices could be dated from the 1976 presidential election, when President Gerald Ford chose nearly to double the wheat loan rate (from $1.37 to $2.25 per bushel) and increase it for corn (from $1.25 to $1.50 per bushel). This unfortunate decision was followed after President Carter's election by further increases written in the 1977 farm bill.[22]

For the 1985 farm bill the aim of President Reagan was to gather a consensus on phasing down of target prices and on calculating the loan rate as 75 percent of the average market prices for the last three years. If it succeeds in doing so, it could become a historical benchmark for American agricultural policy (see Appendix 20).

TRANSFORMATIONS IN FARM OWNERSHIP AND OPERATION

The evolution in American agricultural policy from 1924 to the present is characterized by the adoption of new policy instruments designed to increase the efficiency of government regulatory measures while granting increased freedom to farmers in the management of their operations.

The combination of increasingly sophisticated yet still essentially interventionist programs has achieved its goal by gradually removing government from the marketplace. This result was possible because the adjustment in supply, which was carried out by government programs, was accompanied by a substantial adjustment in the productive structure.

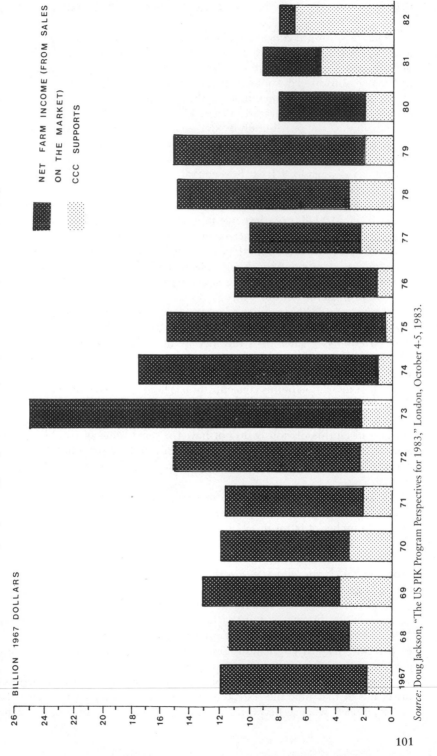

FIGURE 4.9 Farm income declines despite price supports, 1967–1982 (in billions of constant 1967 US dollars)

NET FARM INCOME (FROM SALES ON THE MARKET)

CCC SUPPORTS

BILLION 1967 DOLLARS

Source: Doug Jackson, "The US PIK Program Perspectives for 1983," London, October 4-5, 1983.

The trend toward concentration is easy to illustrate. From 1950 to 1975 there was a reduction of about 50 percent in the total number of farming operations (from 5 to 2.5 million units). At the same time, the amount of land area under cultivation did not vary much, although it followed the evolution in agricultural supply and demand. In all, the average farm size increased from 86 hectares in 1950 to 154 hectares in 1974.

Changes in the distribution by size are also significant. Two precautions must be taken, however, before discussing these figures. First, the income figures given often do not include off-farm income. The resulting problem is illustrated in Appendix 9. It can be seen that (1) the greater the sales figures, the lower the percentage of off-farm income in total income in 1982, 4.5 percent for farms with sales above $500,000, 65.8 percent for farms with sales between $40,000 and $100,000; (2) the percentage of off-farm income rose rapidly and, beginning in 1964, exceeded net farm income; (3) farms with sales figures below $100,000 derived more than two-thirds of their income in 1982 from nonagricultural activities; those with sales above $100,000 less than one-third; and (4) since 1980, the 1.75 million farms with sales below $40,000 have had negative or no net agricultural income.

These last two observations are particularly surprising. They result in the paradox that the total income (net farm and off-farm) of a farmer with sales less than $10,000 is greater than the income of a farmer with sales between $10,000 and $100,000, a phenomenon that explains the growth in part-time work.

A first indication of this concentration is land area. The number of farms that cultivate more than 200 hectares has increased from 6 percent in 1940 to 19 percent in 1970. This indication, however, is not very useful because 100 hectares in Florida obviously do not have the same value as 100 hectares in Kansas. It is therefore preferable to examine the trend toward concentration by considering sales figures. Appendix 10 emphasizes essentially the following points:

1. Since 1959 the number of "commercial" farms (with sales greater than $2,500) has exceeded that of small farms, which represented only 23 percent of the total number of farms in 1979, compared with 59 percent in 1949.

2. The proportion of farms with sales greater than $10,000 increased from 10 percent of all farms in 1949 to 50 percent in 1979. The percentage of "commercial" farms with sales less than $10,000, however, has remained stable, falling only from 32 to 27 percent.

3. There is a strong movement toward the concentration of American agricultural production on large farms with sales greater than $100,000. In 1982, these farms, although representing only 9 percent of all farms, generated 63 percent of all gross agricultural income and 95 percent of net agricultural income.

Beginning in 1973, the world price greatly exceeded the domestic support price. But at that time the restructuring appeared complete or at least stabilized. Indeed, whereas between 1960 and 1970 an average of 100,000 farms closed operations each year, between 1970 and 1974 only 38,500 did so. The following three years, this figure fell to 28,000 per year and from 1978 to 1982 to 9,000 per year. In 1982 there were only 2.4 million farms left in the United States (see Appendix 10), of which 1.7 million had negative or no net agricultural income.[23]

In addition, each year between 1960 and 1970 the agricultural population declined by 4.8 percent. Since 1971, this rate has been below 0.8 percent. This stabilization was possible because the largest production units played an increasingly important role. Until 1970, a strong concentration movement took place. The sudden increase in prices in 1973 was profitable mainly for the largest producers. One might speculate that the government decided that after that date most adjustments would be made by altering the intensity of production by the largest producers. Once this restructuring was believed to have been completed, authorities decided that it was time to play the trade card.

PROTECTIONIST MEASURES

How, then, is it possible to explain the importance American agricultural policy places on protectionist measures? Several observers consider these measures to be the source of all the problems suffered by European products, which are discriminated against in the United States by malevolent federal agencies whose sole aim is to halt the introduction of French cuisine in a country devoted to hot dogs and hamburgers. Others believe protectionism, even for agricultural products, has no place in the home of the free market: U.S. authorities were forced to adopt protectionist measures by the hateful government subsidies, which the EEC, under its common agricultural policy, grants to its food and agricultural exports so as to invade the open market of 220 million U.S. consumers.

In reality, both of these views are erroneous. Agricultural protectionism is a historic pillar for the U.S. agricultural economy. Unlike in Europe, where the creation of the Common Market in the 1960s entailed a total revision of the system of barriers erected beginning with Jules Meline, the protectionist French minister of agriculture of the 1890s, in the United States there is a legislative and sociological continuity that now allows American agricultural policy simultaneously to embrace both free market and protectionist principles depending on the laws used and the products concerned. The most visible manifestations of protectionism must be analyzed on the basis of this observation.

How should the degree of protectionism be determined? A rough evaluation can be made by calculating the difference between the national price and the world price explainable by customs and nontariff barriers established by governments. This formula is simply as follows:[24]

$$\text{Nominal Rate of Protection} = \frac{\text{Domestic Price} - \text{World Price}}{\text{World Price}}$$

This formula does not, however, take into account food and agricultural products delivered on the market whose production required the use of many intermediate products, which can also be protected and thus be delivered at a price above the world price. Chemical products necessary in agriculture, animal feed used by ranchers, or processing treatment and packaging equipment fall into this category. This reasoning led to the concept of real protection:

$$\text{Real Rate of Protection} = \frac{\text{UAV at domestic prices} - \text{UAV at world prices}}{\text{UAV at world prices}}$$

in which UAV is the unit added value, that is, the difference between the value of final production per metric ton produced and the value of intermediate products consumed during the production of this final metric ton.

The figures cited in Table 4.3 are for 1968 but are still indicative of the high degree of protection enjoyed in the United States by a certain number of agricultural sectors, such as tobacco, peanuts, sugar, cotton, and, above all, dairy products. To take the example of dairy products, which will be examined throughout the following discussion, the effect of nontariff barriers is of primary importance: because customs duties are low, about 10 percent on most

TABLE 4.3. Degree of protection of certain products on the U.S. market in 1968 (in percent)

	Normal Rate of Protection	Real Rate of Protection	Tariff Barrier[a]	Nontariff Barrier[a]
Very well-protected products				
Dairy products	16.8	48.2	−3.4	51.6
Tobacco	17.0	28.2	24.5	3.7
Cotton	0.3	100.8	−1.3	102.1
Sugar	195.8	662.2	72.2	590.0
Peanuts	69.3	204.0	[b]	[b]
Not very well-protected products				
Corn	0.0	20.0	[b]	[b]
Soybeans	0.0	−6.7	[b]	[b]
Beef	7.5	13.8	13.2	0.6
Poultry and eggs	0.8	−19.6	−16.6	−3.0

Source: G. Johnson and J. Schnittker, eds., *U.S. Agriculture in a World Context: Policies and Approaches for the Next Decade* (New York: Praeger, 1974).

[a]The real rate of protection is the sum of the tariff barrier (derived from customs duties) and the nontariff barrier (quota, sanitary prohibition).

[b]Not available.

cheeses, import quotas and sanitary requirements have played the most important role.

For dairy products, the sanitary barriers are without doubt the most effective form of protectionism, although the most visible and criticized are the import quotas established under Section 22 of the Agricultural Trade Adjustment Act of 1933. In addition, the Trade Act of 1974 added formidable tools of an obligatory and automatic nature. The measures taken at the end of the Tokyo Round in 1979, however, attenuated their effect.

An examination of the technical conditions for American milk production will permit a better perspective on American protectionism, notwithstanding the free trade rhetoric that normally characterizes U.S. official statements.

Sanitary Protectionism

This form of protectionism is carried out by two agencies: the Department of Agriculture, which jealously protects American livestock from contamination by foot and mouth disease and other Euro-African "plagues," and the Food and Drug Administration, a branch of the Department of Health and Human Services, which, fortunately for French, Chinese, or "ethnic" gastronomy, supervises human health in a more good-natured fashion.

The mechanism established by the animal health services is all the more implacable in that it is founded on objective scientific information, which is admitted by the entire international veterinary community; it divides the world into two categories of countries: (a) totally unaffected countries, that is, FMD-free countries, where vaccinations are not carried out against foot and mouth disease and where procedures against germ-contaminated imports are the same as those used in the United States; and (b) countries that are considered contaminated, that is, all countries other than those in category *a*, without distinction between those that conduct vaccination programs to eliminate all possible sources of contamination and those that truly are populated with sick animals.

Between these two categories there is no middle ground. All links are totally severed, not only with respect to live animals (with the exception of the equine species), but also for fresh and frozen meat (excluding preserved meat), frozen sperm for artificial insemination, and livestock feed. Furthermore, recent scientific concerns have added to the list of restricted products: casein, powdered milk for animal feed, and lactoserum can no longer enter the FMD-free sanctuary. Foot and mouth disease has thus become a problem of world scope.

U.S. sanitary requirements project an image internationally, particularly in Europe, which is quite different from the free market ideas officially espoused by all past American administrations. The requirements can be justified by the high cost of vaccinating 120 million head of cattle semiannually. But they also allow the government to maintain the political support of influential American cattle interests.

Sanitary restrictions also prohibit the import of poultry products (because of Newcastle disease), salted pork products (because of hog cholera and vesicular swine disease), and fruits (cold treatment is required, which would make prices prohibitive). Although these barriers are not well known, they are extremely effective, comparable, in fact, to import quotas, which, as we will now see, have played a fundamental role in shaping U.S. dairy production.

Dairy Import Quotas and the Double Price

These quotas can be established under Section 22 of the Agricultural Adjustment Act of 1933 after a public investigation conducted by the International Trade Commission. This procedure was designed to show that imports are in contradiction with the system of price and quantity controls established by American law for certain products. Section 22 has been used primarily for dairy products, since the professional organization is powerful and efficient: the price of milk intended for consumption as a beverage is set at a much higher level (about 30 percent) than the guaranteed prices in Europe. This high price is maintained through a very strict control on production, which assigns each dairy farmer a production quota; if he exceeds it in certain seasons he will be fined, often substantially, by the local cooperative.

For this reason, import quotas on powdered milk, butter, and eventually almost all cheeses were imposed under Section 22. They are set at a so-called "historic" level, that is, they are calculated based on imports from each country and by each American importer during a reference period decided on by the International Trade Commission. For example, the reference period for blue cheeses was set as the years 1949–50. For France, still recuperating from World War II, this meant deprivation of any import quota. It was entitled to only 5 metric tons per year, although with 30,000 metric tons, France is the world's largest producer of blue cheese produced from cow's milk.

In 1969, at the request of the European Community, the Tariff Commission (now the International Trade Commission) created another quota applied to cheeses priced lower than a set threshold price indexed on the intervention price of milk in the United States. More expensive specialties such as brie and camembert were thus freed from quotas. Revocation of this privilege was decided when the Trade Act of 1979, enacted following conclusion of the GATT Multilateral Trade Negotiations, allowed only some soft ripened cheeses to enter free from quota. Since January 1, 1980, all other cheeses are subject to import quotas.

Section 22 is justified internationally by the GATT provision that authorizes a country to control imports of a product for which it imposes production restrictions on its farmers. It therefore seems difficult for the international community to protest with much vehemence its intensive application to dairy products. Production quotas in the United States, however, apply solely to high-quality milk intended for consumption as a beverage. So-called industrial milks, for which producers are paid much less, are produced in unlimited quantity, with surpluses purchased by the CCC in the form of nonfat dry milk, butter, or cheddar cheese.

It is interesting to compare the price structure for milk in the United States with that in the EEC. The following paradox can be seen: American dairy farmers receive an average price of 6 percent more than their French counterparts for the milk they sell, yet cheese, butter, and nonfat dry milk processors pay 8 percent less than their European competitors for the milk they purchase. The reason lies with the American consumer, who agrees to pay 26 percent more for the pasteurized milk he drinks. This behavior results in a series of paradoxes that have profoundly influenced the American government.

American policy with respect to grain and soybeans has aimed consistently to open trade channels. The contrast with the policy for dairy products is striking because the latter has served to isolate U.S. dairy producers from the rest of the world. Milk imports during the 1970s represented less than 2 percent of production (1.33 percent in 1978), and exports, usually in the form of food assistance, barely exceeded 1 percent of production.

Even so, since 1980 milk support expenditures in the United States have increased steadily, reaching $2.6 billion in 1982. Public stocks amounted to $3.4 billion, representing more than one year of nonfat dry milk production and more than seven months of cheddar cheese production. Thus on November 29, 1983, an act to reduce dairy production was passed, including a decrease in the support price and subsidies to producers who agreed voluntarily to reduce their production by 5 to 30 percent. These minor changes, much less significant than the production quota measures adopted by the EEC in April 1984, demonstrate the power of the American dairy lobby. The November 1983 program, however, resulted in a 4 percent decrease from the 1983 record production (63.5 Mt) and a similar reduction in the number of dairy cows. Nonetheless, the United States is the world's third largest producer of milk, behind the Soviet Union and the EEC, with a specialized herd of 11 million Holstein milk cows.

For these reasons it is useful briefly to examine the technical aspects of milk production in the United States as well as consumption habits. This analysis will assist in explaining the protectionist policies of the United States, which constitute a substantial limitation on the maneuverability of the U.S. government during international negotiations.

The Technical Aspects of Milk Production in the United States

The United States does not appear to suffer from a lack of technical progress in milk production. Achievements have been considerable, as indicated by the constant increase in the production of milk per cow, which was 3.3 percent per year from 1960 to 1972, dropped to zero between 1973 and 1975, but was restored to 2.3 percent between 1976 and 1982. In 1982 yield was 5,600 kilograms of milk for each of the 11 million milk cows counted.

To maintain the productivity of U.S. dairy cattle, farmers have created a breeding stock purely for dairy production: the black and white, also called American Holstein because of its distant European origins. There are more than a thousand dairy herd improvement associations employing two thousand two hundred permanent inspectors in 1977 and including thirty-three thousand of the three-hundred thousand dairy producers in the United States. This top-class breed numbers on the average seventy-six milk cows per livestock operation, or 2.5 million cows in all, nearly one milk cow in four. Their milk yield surpasses the average American level by 25 percent (50 percent in 1960). Forty cooperative or private organizations purchase 11 million doses of Holstein selected bull sperm, which indicates that artificial insemination is used for 70 percent of all dairy cattle.

Disease control has almost totally eliminated bovine tuberculosis and brucellosis. To eradicate brucellosis, the milk of more than seven-hundred thousand herds is analyzed several times per year using the "ring test," which can detect milk from a single diseased animal in a twenty thousand–liter tank. More than 14 million milk cattle sent to the slaughterhouse are diagnosed for brucellosis through blood analysis. Through use of these permanent control systems, more than 7 million blood tests were conducted in the 1970s on the cattle found suspect by the two preceding methods, which made it possible to identify 180,000 animals with brucellosis (less than 1 percent of the total herd) and to vaccinate 3.8 million milk cows and heifers that may have come in contact with the diseased animals. This description of the method for combating brucellosis illustrates the breadth of resources used in the United States for the sanitary protection of cattle. These technical details are also useful in outlining the production context within which American dairy policy has developed and reconfirms the importance and the seriousness of sanitary protection efforts.

The Technical Conditions of Milk Consumption in the United States

Why, then, has the United States not taken advantage of its exceptional situation to export its dairy products throughout the world? This behavior, which is paradoxical for a country in which exports have become vitally imperative, can be justified by the orientation of milk consumption in the United States. More than 50 percent of consumption from 1950 to 1960 (and more than 40 percent currently) involves a nonexportable, perishable product: high-quality pasteurized milk, "grade A." The priority of producing a drinkable milk led the industry to neglect the potential of dairy foods.

Europe, on the other hand, and particularly France, traditionally has had neither the high bacteriological quality drinkable milk nor a refrigerated distribution system. Consequently, the Europeans have oriented their dairy production toward yogurt and more than four hundred types of high-quality cheeses, which now are exported throughout the world with increasing popularity.[25]

For the American dairy producer, cheese is a simple, relatively undiversified by-product. More than 60 percent of American cheese, excluding fresh cheeses, consists of a single variety, the American cheddar. Per capita consumption does not exceed 8 kilograms per year in the United States, as compared with more than 18 kilograms in France. The same is true of butter: the average

American consumes 2 kilograms per year, while the average Frenchman and the average Irishman hold the world's record at 10 kilograms per year.

The policy of producing high-quality milk is the key to the American dairy strategy. In 1933, the Agricultural Adjustment Act, confirmed in 1937 by the Agricultural Marketing Agreement Act, created professional organizations endowed with extensive powers to manage the production and distribution of grade A milk. These forty-seven organizations, the Federal Milk Marketing Orders (FMMO), are similar to the agricultural economic committees created in France by the initial agricultural orientation laws of 1960 and 1962. They have succeeded in obtaining much higher prices for drinking milk than for so-called processing milk (+45 percent in 1964–66, +26 percent in 1974–76, but only +19 percent in 1979–81).[26]

The success of these FMMOs has been so great that their membership has expanded to 110,000 dairy farmers out of a total of 300,000, covering about 65 percent of total production. In fact, in 1978 grade A milk, which qualifies for consumption as a beverage, represented 83 percent of the total collection of 55 Mt, of which only 44 percent was used in beverage form (103 kilograms per inhabitant out of a total per capita consumption of 256 kilograms).

As a result, 50 percent of the milk collected by the FMMOs is downgraded as processing milk, even though its bacteriological qualities would ordinarily permit its consumption as a beverage. Only states such as Iowa, Idaho, Nebraska, the Dakotas, and Minnesota continue to classify less than 60 percent of their production in grade A form. In Wisconsin, the largest dairy state with 18 percent of national production, 68 percent of production was grade A in 1982. The failure to adapt production to demand has handicapped the dairy sector and explains its tendency toward self-sufficiency.

It would be absurd to try to explain the American protectionist movement solely on the basis of the dairy sector. Nor should the problems of other sectors of the economy such as textiles, shoes, and steel be downplayed. Our purpose here will be to show the leadership role of the dairy industry in this movement, given its advanced technical support and formidable economic power based on production organization dating back to 1937.

The formulation of the law which in 1974 authorized President Nixon to negotiate the new round of GATT discussions was a startling demonstration of the central role played by the dairy industry in the elaboration of U.S. trade policy.

The Protectionist Tools of the Trade Act of 1974

The Trade Act of 1974 provided an opportunity for the dairy lobby to endow the United States with new, more efficient protectionist tools. Adopted during the Watergate scandal, the Trade Act of 1974 was the first trade law requiring the executive branch to apply certain protectionist measures, such as countervailing duties, at the simple request of interested American parties. This reinforcement of a legislation already contrary to the general provisions of the

1945 GATT was nonetheless applied. The grandfather clause, or waiver, allowed countries such as the United States that had established trade legislation before 1945 to maintain the principal elements of that legislation. Strengthening of these elements, based on contested interpretations of the GATT procedures, sparked concern in European states and contributed to their more united consciousness in the face of the American threat.

This legal innovation had a number of immediate consequences. In 1975, complaints from American companies over the restitutions allocated by the European Economic Community to their cheese exporters forced the American government to recognize the existence of an export subsidy. According to the new 1974 law, this should have led to the automatic application of a countervailing duty equal in value to the restitution. But, in fact, a special provision of the 1974 Trade Act provided the United States government with a negotiating chip during the Tokyo Round that allowed the government to waive application of this duty for four years if the adverse party agreed to reduce the export subsidy or limit its application. This waiver was applied to cheeses from the EEC, Switzerland, Norway, and other countries and hams from the Netherlands and Denmark. In exchange, export subsidies had to be reduced to varying degrees according to the retaliatory power of each of the countries concerned. Thus Switzerland, which at the time was also negotiating the purchase of supersonic fighter planes, was forced to reduce its export subsidies for emmenthal cheese, and the EEC was forced to eliminate them completely.

Another creation of the Trade Act of 1974 is Section 301, which authorizes economic retaliation against a country that through export subsidies is competing with U.S. exports to a third country. Numerous investigations have been conducted under this provision, but because of the vagueness of the measures authorized precise decisions have not been made.

An evolution in administrative practice has therefore been observed since 1975. The Treasury Department, which up to this time had successfully fought against the protectionist tendencies of the Department of Agriculture, changed its position. The Office of the Special Trade Representative (STR) pushed for the strictest application of the provisions of the Trade Act with the dual objective of (1) controlling the protectionist tendencies within Congress by showing that the executive branch was prepared to limit abusive imports; and (2) enhancing the government's negotiating position in the GATT discussions by creating negotiating chips.

The evolution in French exports illustrates the effectiveness of protectionist measures in the dairy sector vis-à-vis one of the world's largest exporters of dairy products. Exports of milk powder and butter to the United States were totally eliminated by 1975, and exports of casein and emmenthal cheese fell to tiny quantities. The tools used to eliminate these exports were Section 22 for butter, milk powder, and cheddar cheese; threat of countervailing duties if restitutions were maintained for emmenthal; and health regulations for casein (as well as butter and milk powder). Some products have been blocked by two or three separate regulations, to provide better assurance of success. Similarly, it could be shown that the quotas imposed by the Meat Import Act of 1964 limit meat imports to about 7 percent of U.S. consumption and are complemented by health restrictions associated with protection against foot and mouth disease.

The results of the GATT multilateral trade negotiations, incorporated into U.S. law by Congress in July of 1979, do not modify the protected status of the American dairy industry. Of course, the import quotas imposed on the EEC cheeses were increased a little, but high-priced cheeses are no longer excluded from import quotas. The basic protectionist tools available to the American dairy industry were little changed, with the exception of countervailing duties on subsidized imports, which since 1979 can be applied only if it is proved that material injury has resulted.

Significant change in the American dairy industry's status, therefore, appears a distant prospect. This sector will probably remain isolated from the world market for many years to come.

To correct any impression of incoherence the reader might derive from the coexistence of free trade rhetoric and protectionist practices in certain sectors, it is useful to recall the collective nature of American government, which often results in contradictory decisions by the legislative and executive branches.

These contradictory policies appear to be particularly dangerous in the agricultural sector because they may reduce the credibility of American statements favoring the elimination of agricultural and food trade barriers. The American approach appears more coherent when one realizes that the free trade strategy, applied in the grain and soybean sectors as well as to a lesser degree in sugar and wine, requires the political support of cattlemen and a large dairy sector.

Furthermore, these protectionist policies are accepted by American firms involved in trade; they have long understood that trade barriers are not harmful to their interests. On the contrary, their foreign investments enable them to take advantage of trade barriers established by certain countries in retaliation against American protectionism. Accustomed to penetrating foreign borders, free trade proponents can easily tolerate agricultural protectionism by the United States provided that it affects only certain minor sectors of the American economy and that it is sufficiently limited and well defined as to prevent massive retaliation by the countries concerned.

Paradoxically, agricultural protectionism, which serves as both a target and a weapon in international negotiations, could eventually lead to greater foreign receptiveness to the principal American agricultural products. The final results of the GATT multilateral trade negotiations in Geneva from 1969 to 1979 must be viewed from this perspective. These results will be examined in the conclusion of this book, along with the various attempts made by the Reagan administration to open the Japanese market and to prevent reform of the EEC common agricultural policy from hindering access for American products.

The Agriculture Industry

In examining American agricultural policy, we should not overlook the role played by a number of new entrants on the scene. The influence of the political process on the form of intervention has long been evident in the American agricultural sector.

Increasingly, new participants have entered the debate. First are consumer groups, whose objectives are to stabilize prices and assure abundant supplies. Within the past several years, a new group has emerged, known as agribusiness. This group also desires abundant supplies, but its main hope is to achieve the opening of new markets. Under the influence of this group, therefore, trade constraints have become a more vital issue.

It is evident that the multiplication of participants does not facilitate achieving a consensus on agricultural objectives. Programs to control supplies elicit a conflicting response from each political group. Farmers are entirely in favor of programs to place cropland in reserve, consumers will be moderately opposed, and agribusiness will be resolutely hostile. Inversely, farmers are not favorably disposed to a program of reserve stocks, but consumers and agribusiness enthusiastically support it. Its cost, however, is a matter of concern to taxpayers.

A new concept must therefore be introduced, which encompasses farmers who work the land, suppliers of inputs and equipment necessary to agricultural production, and processors and distributors of farm production. This economic complex, known in the United States as "agricultural industry," involves almost one-fourth of the active population of the country. Table 5.1 shows a strong trend in the distribution of employment in this complex: in 1985 agriculture employed 3 percent of the active population, and in 1946 one worker in six was directly involved in agricultural production. In 1985, for each farmer, there were more than five workers upstream and downstream involved in furnishing farm production supplies or equipment, in processing, transporting, distributing, or exporting his products. The agricultural industry is thus the largest industrial sector in the United States.

This situation is not unique to the United States. During a symposium

TABLE 5.1. Distribution of labor in the agriculture industry (millions of workers)

	1946	1966	Percentage of Annual Variation	Projections 1985	2000
Employment in the production of industrial products for agriculture	5.0	5.7	+0.7	6	7
Employment in agriculture	10.0	5.6	−2.9	3	2
Employment in the transformation and the distribution of food, feed, and fibers	9.5	12.0	+1.2	14	17
Total employment in the agriculture industry	24.5	23.3	—	23	26
Total U.S. working population	60.0	76.6	+1.2	96	114
Percentage of the agriculture industry in the total working population	41.0%	30.4%	—	24%	23%

Source: For the years 1946 and 1966, R. Goldberg, *Agribusiness Coordination* (Cambridge, Mass.: Harvard University Press, 1968). Projections for 1985 and 2000 made on the basis of the growth rates from 1946 to 1966.

jointly organized by the FNSEA (the French farmers' union federation) and the CNPF (the French industrialists' union), in June 1979, the figure of 5 to 6 million workers was advanced for the agricultural industry in France, or about 25 percent of the total active French population, whereas agriculture itself employed only 9 percent. Studies in 1981 showed that the French agricultural industry was estimated to account for about 22 percent of the French work force, of which only 8 percent worked in agricultural production, 1 percent upstream, 1 percent in agricultural services, and 12 percent downstream.[1]

The traditional view continues, however, to show agriculture as an activity isolated from the rest of the nation. It is now necessary to change this narrow-minded approach radically and be aware of the new agricultural participants. Those upstream include producers of farm machinery, fertilizer, pesticides, and veterinary products; and producers of livestock feed, calves, seed, and other products consumed in agricultural production. Those downstream include processing industries; storers, transporters, and exporters; and national distributors (supermarkets and fast-food restaurants). Europeans would have difficulty imagining the colossal importance of these industrial giants upstream and downstream of agricultural production in the United States. Nonetheless, farmers continue to play an essential leadership role within this agro-industrial complex.

To confront these new participants and reinforce their negotiating positions, farmers have organized in cooperatives and marketing orders. The industrial groups have themselves attempted to become agricultural producers. This was the strategy from 1955 to 1965. Then, when their efforts resulted in problems and waste, they planned agreements with the cooperatives. This was the strategy from 1965 to 1975. Recent trends are more contradictory. This complex interaction will be studied after the new industrial participants have been examined.

We hope the reader will pardon the following listing, which for some may appear overly detailed. It seemed indispensable to mention at least the principal names in each of the sectors and note their relationships.

THE INDUSTRIAL CORPORATIONS IN FOOD AND AGRICULTURE

In 1982, the farm machinery sector had sales of $11 billion, $8 billion in the United States and $3 billion in exports. Half of these sales were in tractors, one-sixth in combines. Six large multinational firms dominate the American and international market: International Harvester, John Deere (with $4 billion in 1983 sales), Massey-Ferguson, Allis-Chalmers (Fiat Group), White Motors, and Case (Tenneco Group). All of these firms have numerous factories overseas, particularly in Europe. Such is the case for the Canadian group Massey-Ferguson, which before its financial difficulties was the largest exporter of agricultural machines from France. It is also true of John Deere, founded in the small town of Moline, Illinois, which revived three plants in France and became the second largest agricultural equipment exporter from France.

The pesticide sector also occupies a dominant world position.[2] Out of total world sales of $7 billion in 1982, the United States produced $5 billion and exported one-fourth of its production. Twelve companies account for 76 percent of world production. Among these, six are American, controlling one-third of the production. Herbicides represent half of American production in this sector, insecticides 29 percent, and special preparations for the public, 11 percent. The United States also has the world's largest vaccine industry, including such companies as Merck, Sharp and Dhom, which is followed in size by the West German company Hoechst and the French company Merieux-Pasteur.

The fertilizer sector is more complex. The United States has become a permanent importer of nitrogen fertilizer, as a result of compensation agreements with the Soviet Union, where the United States has constructed twelve ammonia plants, and because of the increasing cost of natural gas, which makes plants in the Netherlands and the Middle East more competitive.

The situation is the reverse in phosphate fertilizers. The United States exports one-third of its production, but the world market is saturated. Here again, barter agreements, exchanging superphosphates for ammonia, were planned with the Soviet Union before the embargo of January 1980.

The market for potash fertilizers is dominated by Canada, from which the

United States imports 70 percent of its consumption. Attempts to barter phosphates for potash have been considered with the Soviet Union to reduce American dependence on the Canadian province of Saskatchewan when it decided to nationalize 50 percent of its production.

The relationship between farming and downstream industries is more delicate. Sales by the agricultural sector amounted to $167 billion in 1981,[3] while those of the food industry were $276 billion (50 and 84 respectively in 1967). With 1.7 million workers, the food industry constitutes the largest American industry with about 13 percent of the gross national product in 1978.[4] This industry is the principal market for agricultural production. Its concentration in the hands of a few large corporations, each dominating segments of the market, further reinforces the industry and allows it to exercise a world strategy that can be interpreted not only as an additional lever in its negotiations with the agricultural sector but also as an opportunity for farmers to sell their products on a large scale.[5]

Appendix 11 summarizes the principal sectors of the American food industry: meat, grain processing, dairy products, canned and frozen products, oils and oil meals, alcoholic beverages, soft drinks, coffee, and sugar and derivatives.

The distribution of meat is conducted by the meat packers, a term which encompasses both the slaughter of livestock and the preparation of meat cuts. The three large traditional meat-processing companies, Swift, Armour, and Wilson, have been forced to diversify because their slaughterhouses, which were located far from the production areas, have met competition from new entrants—Iowa Beef Processors, which as subsidiary of Occidental Petroleum has become the largest meat-packing company, Oscar Mayer, and Hygrade Food—which have enjoyed exceptionally rapid growth because of the creation of medium-sized, highly mechanized slaughterhouses located near production areas. The three traditional giants, which accounted for 60 percent of the industry's assets in 1948, accounted for only 30 percent by 1964. Under new management during the 1970s, they closed their slaughterhouses, famous in Chicago and other large cities, and diversified. In 1975, of Swift's $3 billion in sales, only one-third was produced from meat. Meanwhile, two corporations with sales of over $5 billion and $4 billion respectively, the Greyhound Corporation and Esmark, made strong entries into the meat business. Esmark became the owner of Swift-Hunt-Wesson, which he sold to Beatrice Food in 1984.

In the baking industry, the trend has also been away from concentration. The five largest companies, National Biscuits, Continental Baking (which belongs to ITT), Sunshine Biscuits, Campbell-Taggart, and American Bakeries, now represent only one-third of the industry's total assets. Campbell-Taggart enjoys unrivaled technological leadership in the area of industrial baking.

The canning industry is rapidly being concentrated within six companies: Campbell Soup, California Packing, Heinz (which derives half of its sales overseas), Hunt Foods, Libby, and Del Monte (a subsidiary of Reynolds Tobacco). Nonetheless, there has also been rapid growth in this sector among cooperatives such as Tri-Valley Growers.

Livestock feed is produced by enormous companies: Ralston Purina, with $3.8 billion in sales (its French subsidiary, Sanders, was purchased by Generale Occidentale and then by EMC, the French national mining and chemical company), General Mills, and Pillsbury.

The processing of grain into breakfast cereals is dominated by Quaker Oats and Kellogg.

Although somewhat apart from the grain-oilseed-livestock chain, the beverage industry should also be mentioned, since the firms in this sector are rapidly diversifying and expanding internationally. With respect to soft drinks, a distinction must be drawn between, on the one hand, companies that franchise their product worldwide, such as the Coca-Cola Company ($3.6 billion in sales), Pepsico ($3.5 billion), Royal Crown Cola ($350 million), Seven-Up ($250 million), and Dr. Pepper ($225 million) and, on the other hand, companies that bottle franchised beverages, such as General Cinema, producer of such films as *Close Encounters of the Third Kind* and *Saturday Night Fever.* Over the last ten years, General Cinema has acquired twenty-five plants producing Pepsi-Cola, Seven-Up, and Dr. Pepper. Half of its sales of $500 million now derive from soft drinks.

American beer, produced from barley, and other malt beverages are undergoing continual concentration. In 1976, the five largest producers accounted for 68 percent of production: Anheuser-Busch ($1.8 billion in sales), Miller (subsidiary of Philip Morris), Schlitz ($1.16 billion), Pabst ($700 million), and Coors ($600 million). Coors owns its own malthouse, its own can factory, trucks, engineering office, and advertising agent.

With respect to the spirits industry, beginning from the distillation of corn, many of the giants established themselves in Canada because of Prohibition: National Distillers ($1.6 billion in 1977 spirits sales, about one-third of its total sales, with the remainder derived from chemicals), Seagram ($1.2 billion), Heublein ($800 million), Brown-Forman ($330 million), Hiram Walker, owner of Canadian Club and Cognac Courvoisier ($380 million), and Publicker ($230 million) are the six largest producers and account for 50 percent of the industry's production of alcoholic beverages.

The wine industry is dominated by Gallo (the world's largest producer with 1 billion liters in storage capacity and $12 million in publicity); United Vintners, a subsidiary of Heublein and California cooperative, Allied Grape Growers (400 million liters); Guild (200 million liters); Vie-Del (150 million liters); Bear Mountain, a subsidiary of Coca-Cola of New York (120 million liters); Paul Masson, a subsidiary of Seagram (120 million liters); Christian Brothers (100 million liters); and Almaden, a subsidiary of National Distillers (100 million liters).

Tobacco is the fifth largest agricultural product in the United States, with $1.8 billion exported and $.4 billion imported. Tobacco processing is dominated by five large companies: Reynolds ($6.5 billion in 1977 sales), Philip Morris ($5.2 billion), American Brands ($4.6 billion), Lorillard ($3.2 billion), and Ligget ($1 billion). Confronted with an inevitable decline in consumption, they are diversifying rapidly into beer and coffee (Miller-Lowenbrau and General Foods for Philip Morris) or canning, can manufacturing, and biscuits (Del Monte, Sea-Land, and Nabisco for Reynolds).

MERCHANTS

American farmers are thus confronted with about fifty industrial corporations with sales exceeding $1 billion. The costs of processing and distributing food products have remained stable for twenty years and represent 60 percent of the retail price. Furthermore, processing of food used by American families is constantly increasing. It is easy to conclude that distribution, given its prodigious organization, allows American farmers to maintain their share (one-third) of total spending on food ($300 billion spent by consumers in 1982 for domestically produced food—$400 billion including imported products, fish, and self-consumption).

The distribution of food products in the United States is dominated by supermarkets. Of less importance are the hypermarkets (such as the Chicago Carrefour, with more than 100,000 square feet), convenience stores, and gourmet shops, which sell specialties, often imported. Supermarkets are always found in twos, sometimes in threes. Side by side in the same shopping center, they compete intensely, with meat quality, discounts, and promotions constituting the main competitive weapons. The two largest chains each have more than five thousand stores: A and P (Atlantic and Pacific) and Safeway. The Grand Union chain, with more than four hundred supermarkets on the East Coast, now belongs to James Goldsmith's Anglo-French trust, Generale Occidentale.

A typically American phenomenon has come to rival the lifestyle of weekly supermarket purchases: the rapid development of fast-food restaurants— chains serving pizza, hamburgers, and other dishes at competitive prices. In 1965, 22 percent of food expenditures by American families bypassed the family kitchen, going to restaurants, schools, and hospital cafeterias. Twelve years later, this figure had increased to 32 percent[6] (see Table 5.2).

Frozen specialties are developing rapidly for the supply of these "restaurants," which often simply thaw in microwave ovens food prepared in specialized factories ($3 billion in 1977 sales).

Many American conglomerates seek to buy their own fast-food chain. Kentucky Fried Chicken, for example, belongs to Heublein, Pizza Hut and Taco Bell to Pepsico, Burger King to Pillsbury, Burger Chef to General Foods, and Vie de France to Grands Moulins de Paris. Only McDonald's remains independent.

The recession slowed the growth of these chains. Yet, to a large degree, family meals out customarily take place in such establishments, as opposed to

TABLE 5.2. The importance of cafeterias and restaurants in the American family food budget in 1977 (billions of dollars at the retail price)

Billions of Dollars	Total Spending	In Restaurants	Under Collective Ownership	Percentage of Spending Outside the Home
Red meat	52.3	21.7	3.2	45.0
Poultry	13.7	4.2	0.6	35.0
Dairy products	27.4	6.6	2.3	32.4
Fruits and vegetables	38.9	4.5	2.3	17.5
Cereal food products	5.6	0.8	0.3	19.6
Bread and pastries	19.6	5.3	1.3	33.6
Other	26.0	6.3	1.5	30.0
Total	186.4	49.4	11.4	32.5

Source: Jean-Claude Trunel, "The Evolution of Food Prices in the United States," note by the French agricultural attaché to Washington, May 1979.

the European custom of dining in conventional restaurants. Fast-food establishments, therefore, were affected by the 1977 American recession, but not as severely as conventional restaurants.

Just as the United States has developed its distribution of food products for sale within the domestic market, so has it conquered foreign markets. One of the most impressive components of the agriculture industry is the six great grain and oilseed trading companies: Cargill of Minneapolis and its subsidiary, Tradax, in Geneva, Continental of New York and Brussels, Bunge of Buenos Aires and Rotterdam, Dreyfus of Paris, André of Geneva and its subsidiary, Garnac, in New York, and Cook of Memphis. Cook almost disappeared in 1977 following an abnormal rise in the price of soybeans and an unfortunate decision of the "Cook's whiz kids" to sell short when the Bunker Hunt family was buying long.

Between 1970 and 1975, these companies shipped 96 percent of the wheat, 95 percent of the corn, 90 percent of the barley, and 80 percent of the sorghum exported by the United States. With respect to non-American commodities, they exported 80 percent of Argentine wheat, 90 percent of Australian sorghum, and 90 percent of the wheat and corn from the European Community.[7]

Dan Morgan defines these companies as those that know how to operate the pipeline channeling grain from the farmer to the consumer, through their complex system of silos, ships, trains, telex, banks, and, most recently, grain and soybean processing plants.[8] Most of these companies are not of American origin, but all have located their information centers in New York[9] and have set up subsidiaries in Geneva, vital for conducting new transactions sheltered from the declaration requirements imposed by the Export Control Act.

Industrial corporations, such as the petroleum firm Philips Brothers, have also become interested in this sector,[10] as have a number of Japanese corporations (Mitsubishi, Mitsui, Itoh, Marubeni, Somitomo, and the cooperative Zen-Noh), which have succeeded in capturing 20 percent of the American grain export market, estimated at about 130 million metric tons. Only 12 percent currently goes to Japan; the other 8 percent is international trade.

With respect to international trade in fruit, United Brands (formerly United Fruit) controls 35 percent of the world banana market, Castle and Cook controls 25 percent, and Del Monte 10 percent. In 1974 they successfully resisted an attempt by producer countries to create an OPEC for bananas.[11]

In world wine trade, U.S. firms have shown once again their management capabilities, even if primarily in imports to the United States. Among the forty largest brands of wine imported to the United States some names are well known and others will be: Riunite Lambrusco of Italy is imported by the House of Banfi, Mateus of Portugal by Dreyfus-Ashby (Schenley–Rapid American), Yago of Spain by Mr. Henri (Pepsico), Bolla of Italy by Brown-Forman, Lancers of Portugal by Heublein, Folonari of Italy by Foremost-McKesson, and Mouton Cadet of France by Northwest Industries and its subsidiary Buckingham Wine Corporation, lately bought by Beatrice Food and resold to Whitbread.

Commodity futures markets are the indispensable auxiliary to trade and American agricultural policy. Since this policy exposes American cooperatives, dealers, processors, and exporters to fluctuating world market prices, it was necessary to provide them a means of protecting themselves against daily variations in market prices. The role of the commodity futures markets is to allow long-term transactions to be carried out at firm prices. Hedging is their principal function. Since the 1960s, very rapid development has been seen in these futures markets in the United States (37 million contracts in 1976 and 99 million in 1981). The largest markets are in Chicago, at the Board of Trade, which deals in wheat, corn, and soybeans; the Mercantile Exchange, where futures contracts on livestock products and financial or monetary products are traded; the Mid-America Commodities Exchange, which specializes in small contracts; and in New York, the NYMEX, which deals in contracts for potatoes, petroleum products, and platinum; the Cotton Exchange, which deals in cotton and orange juice; and the Coffee Sugar Cocoa Exchange. The various varieties of wheat are quoted at the Kansas City Board of Trade and at the Minneapolis Grain Exchange, the world's largest single wheat market.

To control the operations of these markets, which are private companies for which seats are sold at auction (in June 1982, a seat at the Chicago International Monetary Market sold for $210,000, $11,500 at the Mid-America), a Commodities Exchange Administration was created by a 1936 law. It was replaced in 1974 by the Commodity Futures Trading Commission, which had a budget of $23 million in 1984.

THE CONSUMER ERA

Confronted by these powerful agri-industrial conglomerates, American consumers have progressively found means of control and reaction, which, with varying degrees of success, play an important role. The evolution in food expenditures, however, appears favorable to the interests of the American consumer. The proportion of individual income spent on food has indeed declined from 18.5 percent in 1970 to 16.7 percent in 1977.[12] To make accurate international comparisons (in France, food represented 21.4 percent of household expenditures in 1970 and 18.8 percent in 1977), expenses for alcoholic beverages must be added. In the United States, for moral and historical reasons (Prohibition), purchases of alcohol are not considered part of the food budget. Expenditures for alcoholic beverages in 1977, as mentioned previously, amounted to $11 billion wholesale and $33 billion retail, which would increase household food costs by about 15 percent and increase their percentage of household expenditures by about 2.3 percent, putting them at 19 percent, a level equivalent to that of France the same year.

Content in his dietary habits, the American consumer has become concerned over the trend in retail food prices, which since 1972 have been growing faster than prices in general; this is a reversal of the trend in the past.[13] The position of consumers vis-à-vis farmers should be qualified because the increase in the retail price index for food results as much from processing costs as from agricultural production costs. Indeed, the portion of retail food prices received by farmers has tended to stabilize or decline slightly since 1973.[14] After reaching 45 percent in 1973, their portion declined to 35 percent in 1982. This evolution has been observed for the entire range of food and agricultural products, with the exception of dairy products, poultry, and eggs, for which the farmer's percentage has been maintained.

In the general context of inflation, actions by the federal agencies responsible for protecting consumers are interpreted in different ways. Some observers attribute additional price increases for food staples to the cost of regulatory compliance. Despite the current deregulatory movement, it would be a mistake to assume that consumerism in the United States is subsiding.[15]

The most important of the consumer protection agencies, the Food and Drug Administration, has been involved in approving a series of new products resulting from the development of new technologies (flexible aluminum plastic envelopes for food conservation and polyester bottles) and responding to increased demands from consumer groups (label information on solid food weight for food canned with liquid, a ban on saccharin, prohibition as of September 1977 of plastic bottles containing acrylonitrile, and label information on the origin of animal or vegetable fat used in products, particularly cakes).

The most recent of the federal agencies, the Environmental Protection Agency, has proven an equally formidable opponent to industry, particularly the chemical industry. This agency gained renown in the fight against pesticides by monitoring traces in food and setting contamination limits for sensitive items, such as dairy products, and by regulating or banning their production or regulating their use.

The Bureau of Alcohol, Tobacco and Firearms is responsible for protecting consumers with respect to alcoholic beverages. Its position has been important in several areas, such as metric conversion, with the introduction of the 75 centiliters bottle in 1979 and the reduction in the number of authorized bottle sizes from thirty-two to six; prohibition of plastic bottles for alcoholic beverages; and stricter regulations for the labeling of vineyard wines and wines with appellation of origin.

The fifty states are often stricter regulators than the federal government. This is true of Oregon, Vermont, Maine, and Michigan, which as early as 1980 had adopted bottle return laws, and Hawaii and Massachusetts, which have prohibited the use of aluminum cans for beverages; South Dakota requires biodegradable envelopes.

The Federal Trade Commission has the mission of prosecuting activities in restraint of competition. The case of exclusive territorial franchises for certain soft drinks was recently investigated, suggesting the possibility of a change in the bottling companies' zones of influence, but the Supreme Court decided to maintain the current system, which has been applied for seventy-five years, allowing exclusive distribution in a given zone.

The Better Business Bureau and the Consumer Product Safety Commission assist in resolving disputes between consumers and industry. Their mediation has proven so effective that each cabinet department now has a unit specialized in this area. At the USDA, an assistant secretary, Carol Tucker, was in charge under the Carter administration of all offices involved in quality control and consumer protection. This position was eliminated under President Reagan.

American farmers understand the crucial role played by consumers and their many dynamic associations (*Consumer Reports,* Ralph Nader, and the like) in the orientation and fine tuning of federal regulations affecting agriculture. For this reason, they have taken the initiative of creating a consultative and cooperative organization, the Council of Agriculture of America. This council has conducted studies, investigations, and information campaigns enabling a better understanding between these two population groups, whose interests frequently diverge. These initiatives would be insufficient if the USDA budget were not equitably divided among income support programs for producers and food assistance programs for the economically disfavored and schoolchildren (see Table 5.3).

Thomas Foley, representative from the state of Washington, Democratic chairman of the House Agriculture Committee since 1975, and since 1980 majority whip in the House of Representatives, is convinced that the only means of obtaining votes favorable to farmers from an essentially urban Con-

TABLE 5.3. Distribution of the agriculture budget, 1976–1985 (billions of current U.S. dollars)

	1976	1977	1978	1979	1980	1981	1982	1983	1984	1985
Farmers' income support programs (CCC)	1.9	3.9	5.8	4.4	2.8	4.1	12.0	19.1	7.3	14.4
Foreign aid programs (PL 480)	1.1	1.5	1.6	1.4	0.7	1.1	1.0	1.0	1.1	1.6
Food stamps and school programs	7.5	7.7	8.1	9.1	13.1	14.0	16.1	18.7	18.7	18.6
Total USDA budget	16.7	16.7	20.4	20.6	24.6	26.0	36.2	46.2	34.6	37.7

Source: French Agricultural Attaché Bulletin, Washington, D.C., various dates; and Carol Brookins, "Agricultural Policy," World Perspectives, December 1, 1984.

gress is to provide an increasing share of funds for the improvement of the quality and quantity of food for the American population (Table 5.3 shows this inevitable evolution).

Federal subsidies for the School Lunch Program date from 1943. It has gradually grown to include more than 25 million schoolchildren, half of whom are exempt from any financial contribution. Its cost exceeded $3 billion in 1978, including 20 million pints of milk distributed in schools every day under the special milk program. The Food Stamp Program was created by the Kennedy administration in 1961, when only fifty thousand people were able to benefit from it. In 1976, the program benefited 18.5 million persons. In 1978, 16 million beneficiaries purchased more than $8 billion worth of food, spending only slightly more than $3 billion. The difference, $5 billion, came from federal subsidies paid directly by the government to the supermarkets. In 1983, 21.5 million persons could claim this aid, the cost of which exceeded $11 billion, despite attempts of the Reagan administration to reduce the scope of the program and replace it with a much more limited program for women, infants, and children (the WIC Program).

The size of these programs may appear surprising in this country of abundance, but they increase domestic food consumption by 1 percent,[16] a direct benefit to American farmers in the form of increased domestic demand. The programs also benefited 8 percent of the American population, whose diet was insufficient or unbalanced, and 53 percent of the population of Puerto Rico in 1979.

The size of these food assistance programs invites consideration of their economic significance. Indeed, if a family spends 35 percent of its income for food and if the program allows it to reduce its food expenses to 25 percent, it can use the remaining 10 percent for either food or nonfood needs. An investigation conducted in 1972 and 1974[17] showed that only 55 percent of this saving was spent to improve family nutrition, with the remaining 45 percent spent on satisfying nonfood needs.

The income redistribution effect of these programs is significant. In 1974, for expenditures of $2.7 billion, more than seventy-five thousand jobs were created as a result of this program, particularly in the agricultural sector (thirty thousand), the food sector (fifteen thousand), the wholesale sector (ten thousand), and the retail sector (fifty thousand). Purchases of meat and canned and frozen products were helped the most by these transfers. The Gross National Product grew by $.4 billion, the equivalent of 15 percent of the $2.7 billion transferred. It can be concluded that these aid programs have a marked effect on consumption by the disfavored segments of the population but that their effect on farm income is slight, although not negligible.

THE REACTION AMONG FARMERS

The attitude of farmers confronted with these new participants—the industrial conglomerates and consumer organizations—is indicative of their dynamism. The first effort by the industrial corporations to enter the agricultural sector does not seem to have met with much success, as suggested by the results of the 1974 census. One might suppose that the 2 percent shown in Table 5.4 for the "very large commercial farms" represents large industrial companies, particularly since in 1982 the 25,000 farms with sales exceeding $500,000 earned 28 percent of gross agricultural income and 60 percent of net agricultural income. This, however, does not appear to be the case: in 1978, 52,500, or only 2 percent of the farms were managed by companies and 95 percent of these were family companies constituted to avoid a joint ownership situation at the time of inheritance. Only 2,000 farming companies representing 5 percent of overall farm sales had more than ten stockholders. It must be noted, however, that in California, farming operations that produce sales figures in excess of $500,000, of which many are nonfamily companies, account for 61 percent of total agricultural sales in the state. In 1977, 12 percent of young cattle were fattened in feedlots owned, at least to a great extent, by slaughterhouse companies, the meat packers. In Florida a large number of citrus groves belong to investment firms.

As noted by K. L. Robinson, the evidence is convincing: if giant farming operations had been fully able to use the advantages of scale, they could have little by little displaced farms with one or two workers.[18] Economic reasons played a more important role than government regulations in maintaining family farms. Such regulations are, furthermore, rare in the United States. In fact, most modern equipment is used at optimal level by one or two full-time

TABLE 5.4. Importance of large agricultural operations, 1974

	Number of Farms	Percent of the Total	Percent of the Land Area	Percent of the Sales
Part-time farms (sales of less than $10,000 per year)	1,203,000	52	23	5
Small commercial farms (sales from $10,000 to $39,999)	632,000	27	26	16
Average commercial farms (sales from $40,000 to $99,999 per year)	324,000	14	24	25
Large commercial farms (sales from $100,000 to $199,999 per year)	101,000	5	13	17
Very large commercial farms (sales higher than $200,000 per year)	51,000	2	14	37
Total	2,311,000	100	100	100

Source: General Accounting Office, *Analysis of Agricultural 1974 Census Data, Changing Character and Structure of American Agriculture: An Overview.* See also Appendixes 9 and 10 for 1979 and 1982.

workers willing to work more than ten hours per day during certain periods of the year. This is the perfect description of the family farm, occasionally assisted by two or three neighbors or relatives. These economic factors constitute one reason for the continuation of the family farm.

A second factor that has contributed to the survival of the family structure in American farming is the composite nature of farm family income, already mentioned in another context in Chapter 4 and Appendix 9. Since 1964, with the exception of 1973–75, the off-farm income of American farmers has been greater than their net farm income.[19] In 1982 off-farm income was 80 percent greater than net farm income. Of course, the smaller the farm, the more important off-farm income becomes, which is a factor in the survival of many small farms whose income would otherwise be too small to support a family. Yet, even for the truly commercial farms, with incomes exceeding $40,000 in 1982 (they represented 26 percent numerically and 83 percent of sales), one-fourth of family income came from nonagricultural sources.[20] This off-farm income comes either from salaried work by a member of the family living on the farm but not working on it or from a variety of nonagricultural activities undertaken by the farmers and family assistants: transport, consulting or teaching, representing farm equipment companies, repairs, and so on. This double income has played an important role in slowing, and as of 1970 even reversing, the decline in rural population while preserving a certain number of services. Since 1970, rural population has been growing faster (1.2 percent per year) than urban population (0.7 percent per year).

In contrast, several industrial corporations that purchased land between 1960 and 1970 have since resold it because their profits were insufficient to cover management costs. Ralston Purina, Gates Rubber, and CBK, for example, all made substantial investments in farmland during the 1960s, and all later resold during the 1970s.

Ten states have nonetheless passed legislation limiting farm purchases by companies.[21] These are all large states, such as Texas, Wisconsin, Iowa, Minnesota, Nebraska, Kansas, Missouri, Oklahoma, North Dakota, and South Dakota. In these states, companies own 1 percent of the farms and 4.8 percent of the land area and account for 13.8 percent of sales; in the United States as a whole these figures amount to 1.7 percent, 10.1 percent, and 17.7 percent respectively (1974 census). Legislation in these ten states limited farms managed by companies with more than ten stockholders to 4.2 percent of sales, compared with 5.3 percent for the United States as a whole. These laws have thus had a modest effect, and the problem of corporate farming is for now a minor issue in the United States.

Failing in their attempts at farm ownership, the industrial conglomerates planned an alternative strategy to control American agriculture. This strategy consists of integrating producers by binding them in delivery contracts for their production and by the use of exclusive purchase contracts for their supplies (concentrated feed, one-day chicks, fertilizers, and pesticides). A study conducted in 1972 estimated the proportion of American agricultural production under contract at 17 percent and that produced by corporations at 5 percent.[22] Integration was found to be particularly pronounced in the production of chickens (97 percent), turkeys (54 percent), sugar beets and cane (100 percent), certain vegetables produced for processing (95 percent), citrus fruit (85 percent), and drinking milk (98 percent) (Appendix 12).

The Cooperatives

While corporations have attempted, with varying degrees of success, to penetrate into agricultural production, farmers have taken effective advantage of privileges accorded them by the Capper-Volstead Act. This law permitted the creation of agricultural cooperatives by exempting them from antitrust laws on noncompetitive practices. Through their cooperatives, farmers have been able to enter the processing, distribution, and service sectors.

Supply cooperatives allow farmers to negotiate as a group to obtain better prices from suppliers for livestock feed, petroleum products, pesticides, and seeds. There are more than five thousand supply cooperatives in the United

States, the largest of which are Farmland ($3.2 billion in 1977 sales) and Agway ($1.7 billion).

The processing and marketing cooperatives are powerful in the grain, milk, fruit, and beef sectors. The size of these enterprises is comparable to that of the giants in these sectors. The largest cooperatives in this group include Goldkist ($1.3 billion in sales) in poultry production, Sunkist ($.6 billion in sales) or Florida Citrus Exchange in the production of citrus fruit, or Ampi ($1.8 billion in 1977 sales), Land o Lakes ($1.5 billion), and Dairylea ($.4 billion) in dairy production.

Cooperatives for service, trucking, storage (refrigerated or not), agricultural credit, electricity and rural telephone, mutual insurance, irrigation, artificial insemination, and milk control number nearly seventeen thousand, of which seventy-five hundred are irrigation companies.

Today, farmers demand and play a key role in agribusiness. In 1980 they marketed 35 percent of farm production sold on the market and bought about 13 percent of intermediate products necessary to agricultural production (see Appendix 13). Their position is prominent in the marketing of milk (83 percent), rice (67 percent), and fruits and vegetables (31 percent), as well as in fertilizers (35 percent), pesticides (33 percent), and fuels (25 percent). Their role is less significant in the marketing of beef (18 percent), poultry (25 percent), grain and soybeans (40 percent), as well as in livestock feed supplies (19 percent) and seeds (14 percent).

In addition to cooperatives, marketing orders constitute the second tool used by farmers. They allow farmers, organized in groups, to form contracts with industrial companies. But the bargaining associations for the negotiation of contracts with industrial purchasers are plagued by the existence of minorities of farmers who refuse to comply with collective discipline.

The Agricultural Adjustment Act of 1933, nullified by the Supreme Court and later replaced by the Agricultural Marketing Act of 1937, laid the foundation for procedures allowing the imposition of collective discipline rules on all the farmers of a region. This occurs when a referendum receives a majority of 50 percent of the industrial votes and two-thirds of the farm votes (or of farmers producing two-thirds of the harvest). In these referendums, cooperatives can vote on behalf of all their members.

These marketing orders, directed by an administrator named by the secretary of agriculture, are currently the source of American agricultural policy for citrus fruit and most fruits and vegetables, including hops, potatoes, and tobacco. They cannot, however, be applied to grains, oilseeds, sugar, meats and poultry, potatoes, and apples for processing because these products are covered by federal legislation.

For drinking milk, special marketing orders assign individual production quotas to dairy farmers according to dairy basin (there are about fifty in the United States) and set minimum prices which factories must pay (see the end of Chapter 4 for the dual price system for milk).

Direct price-setting measures are strictly prohibited in the marketing orders, with the exception of those concerning drinking milk. Their activity must be limited to indirect price stabilization actions, including reducing total quantities placed on the market during certain periods by manufacturing or distributing companies, setting individual production quotas for each farmer and quotas for each company limiting the amount it can place on the market, sharing between farmers and companies the cost of disposing of surpluses, in particular the cost of low-price sales to processors and exporters or the financing of stocks, limiting the quantities sold for certain qualities (1947 amendment) or for certain sizes (1954 amendment), inspecting the qualities and prices of products sold in compliance with obligatory prior declaration by the manufacturing or distributing companies, and, finally, excluding the import of foreign products that do not comply with the standards of quality and size defined for the domestic market.

These powers are considerable and have met with harsh criticism from proponents of free enterprise, who associate them with the sterilizing effects of a centrally planned economy. This debate brings to mind similar debates that took place in France before the votes on the Agricultural Orientation Law in 1960 and the law on agricultural offices in 1982 concerning the powers of private and public interprofessional associations.

These interprofessional organizations are effective by allowing farmers to continue drawing an acceptable percentage from retail sales of their fresh or processed products—approximately 30 percent for fresh fruits and vegetables and approximately 50 percent for dairy products—without having to resort to government price guarantee systems, as is the case for grains, oilseeds, cotton, tobacco, and processed milk. Thus they harmoniously complete the instruments of American agricultural policy analyzed in Chapter 4.

The results obtained by farmers assembled in their unions, their cooperatives, and their bargaining associations have thus been favorable to the preservation of their negotiating power. For the entire range of food products, the percentage drawn by farmers from the retail price has reached about 40 percent, varying between 37 and 46 percent in recent years. A decrease to 32 percent occurred in 1979 because of the increase in the cost of labor used in processing and distribution.[23]

These average figures encompass a great disparity between individual products and an undeniable deterioration in the position of farmers between 1972 and 1982 in the following products: meats (from 56 to 51 percent in ten years), edible oils (from 27 to 22 percent), canned fruits and vegetables (from 19 to 17), and bread, bakery, and cereal products (from 17 to 12). The small cooperatives and the family farms now seem to be enjoying a renewed interest from the government in small and medium-sized enterprises, the absence of which has created insurmountable problems in the Eastern bloc and developing countries for the production of spare parts, highly specialized lines of products, and a large number of services necessary for industrial development such as repairs and short distance transport. This desire to reinforce family farms was shown in projects studied by the Carter administration for possible inclusion in the farm bill that was passed in 1981. It corresponds to a sentiment within a segment of the population that a return to small, human units of production will gradually come about, even in the largely industrialized countries. The deregulatory efforts of the Reagan administration share this point of view, although based on opposing philosophical principles.

The large cooperatives, however, are now recognized by the industrial conglomerates as full-fledged partners. An example is the grape producers' cooperative, Allied Grape Growers of Fresno, California, which, with Heublein, created a subsidiary, United Vintners, which now is the number two producer of wine in the United States, behind the Gallo family firm.

The large grain cooperatives, however, participate in only 10 percent of American exports, yet they collect 40 percent of the harvest (see Appendixes 13 and 14). These cooperatives, which own their own port silos, their own livestock feed plants, their own fleets of trucks and ships, and their own petroleum refineries through the International Cooperative Petroleum Associates, do not have their own export network and have always refused to enter this risky area, which has become the specialty of the five large trading companies. The acquisition in 1980 of 50 percent of the capital of the German trading firm Alfred Toepfer by a group of six American cooperatives and four European cooperatives, however, is a sign that the international cooperative movement has become aware of the necessity to be present in world markets. This awareness, however, has come too late to substantially change the position held by the big five traders.

In the final analysis, American agricultural policy is the result of a compromise between the different transactors, each pursuing a particular objective. At the same time, it is an astonishingly dynamic tool. Has this policy been successful? No one believes that it has completely, particularly not the participants themselves. Yet, confronted with the challenge of a constantly evolving economy and the specific nature of agricultural problems, American policy seems to be pursuing the achievement of equilibrium in three areas:

1. Balance in the supply and demand for agricultural products: a policy seems to have been found which though dynamic has avoided the creation of enormous stocks. Just as in 1930, 1960, and 1983, however, the possibility of occasional stockpiling requiring special and costly measures to reduce should not be ruled out.

2. Balance between consumers' needs for food at reasonable prices and farmers' needs for income comparable to that of other social classes: acceptable compromises have been designed that have led to a continuous enlargement of the typical commercial family farm.

3. Balance between small and large farms and between cattlemen (who want inexpensive livestock feed) and producers of grain and oilseeds (who want to export at the highest possible price): their interests are often contradic-

tory, yet each of these segments continues to develop simultaneously in the United States, thanks in particular to the addition of off-farm income for small and medium-sized farms.

This triple equilibrium, itself constantly changing, characterizes American agricultural policy, which for a half century has attempted to create a more flexible agricultural sector, known in the past for its internal and external rigidity, and has succeeded in gradually fashioning what can be called the "American model."

The American Model

The constant adaptation of American policy instruments to the requirements of national and international demand should not lead to the simplistic idea that the technological and financial means of its development have been left in the hands of private enterprise and that government has been willing to relinquish all means of intervention.

The reality is quite different. The American agricultural model was not created by accident. It was created by a constant national will to develop the research, training, educational, and analytical facilities necessary for production and trade. Farm financing has been designed to encourage the creation of a land ownership structure consisting of large, family-managed farms. Financing for the sale of agricultural products always is targeted toward favoring exports and penetrating new markets.

It is, however, indispensable first to show the expansionist strength of the American model, in terms both of crops such as wheat, corn, and soybeans and of livestock through the use of intensive husbandry methods. It is difficult to become number one around the world, yet America's ability to hold that position and increase its lead is truly a challenge to other nations. Before blindly copying this model or criticizing its social and political aspects, it is necessary to examine its performance.

THE TECHNOLOGICAL REVOLUTIONS

The American model could be defined as a series of technologies applied as appropriate for each farming operation, which have endowed agriculture with an unquestioned competitive advantage.

The average productivity of American agriculture, defined as the ratio of output to input, has been increasing constantly at an annual rate of 1.5 percent from 1959 to 1972 and 1.2 percent from 1972 to 1979 (see Appendix 15). The overall productivity growth rate of American agriculture has decreased only moderately despite the energy crisis in 1973 and an increase of 16 percent per year in constant dollars in the price of land from 1975 to 1980 (previously, this rate was only 5 percent per year). The extraordinary increase in the return on labor, which continues at a rate of 5.5 percent per year, is the principal reason for this result, along with the increased return on land (2.6 percent per year). The return on machinery, fertilizer, and pesticides has been diminishing since the beginning of the energy crisis in 1973 by about 6.0 percent per year.

The cost of energy used in agriculture in 1980, however, equaled the value of production on an area of land five times smaller than that required in 1930 to cover energy costs because of the use of draft animals at that earlier time.[1] Technological advances are therefore more costly now than before 1973, but they have not ceased—far from it, because the growth in labor productivity has been maintained at such a rapid pace.

Between 1925 and 1929, it took seventy-six hours to cultivate one hectare of corn; between 1955 and 1959, it took twenty-five hours; today it takes only nine. Since yields have more than tripled in fifty years, each ton of corn requires only one hour and a half of work as compared with forty-five hours before World War II.[2] Similar results are true for wheat, soybeans, and beef.

The Worldwide Dominance of Hard Red and Soft Red Winter Wheat

American wheat is a dilemma for Europeans. Although yields per hectare have indeed increased over the past twenty years from 1.4 to 2.9 tons, they are still barely half of those in the European Economic Community and only slightly more than 50 percent of those in the Soviet Union. On the average in 1979, 1980, and 1981, the yield in France was 4.91 tons per hectare as compared with 2.29 in the United States, 1.55 in the Soviet Union, 1.55 in Argentina, and 1.26 in Australia. And yet, as indicated in Table 6.1, showing the average 1979–81 annual productions, yields, and land areas, the United States is in the list of the world's four largest producers of wheat along with China, the EEC, and the Soviet Union. What is more, with less than 15 percent of world production, the United States has a 40 to 50 percent share of world wheat exports, exporting more than two-thirds of its production. This is explained in large part by the extraordinary reputation of Hard Red Winter (HRW) Wheat, which has a high protein content and is so renowned for its breadmaking qualities that it is considered the ideal wheat with which to mix local, inferior wheats to obtain an improved quality bread.

Sixty-eight percent of HRW wheat grown in the United States is exported, constituting 43 percent of U.S. wheat exports in 1981 (59 percent in 1970). The entire world has become accustomed to the notion that a loaf of bread is not properly made if a sufficient quantity of HRW wheat has not been used in preparing the flour. This notion became so ingrained that Great Britain, for example, was almost totally relying on U.S. HRW for its breadmaking. It took ten years after its signing into the EEC to reach the level at which 70 percent of English flour was made from English wheat, which shows that industrial habits do evolve after all, but very slowly, and that even in Europe farmers know how to produce for the market.

The importance of HRW wheat in American wheat production is ex-

TABLE 6.1. World production of wheat, 1979–1981

	Production (millions of metric tons)	Yield (ton/ha)	Land Area (millions of hectares)
USSR	92.1	1.55	59.5
United States	66.2	2.29	28.9
China	58.0	1.99	29.1
India	34.6	1.55	22.3
France	22.0	4.89	4.5
Canada	20.3	1.80	11.3
Turkey	17.1	1.84	9.3
Australia	14.5	1.26	11.5
Pakistan	10.7	1.57	5.8
Italy	9.0	2.65	3.4
West Germany	8.2	4.97	1.6
Argentina	8.0	1.55	5.2
United Kingdom	8.0	5.60	1.4
World	443.9	1.89	235.1

Source: FAO, Production Yearbook, 1981.
Note: World wheat production reached 523 million tons in the 1984–85 record year, of which 75 were for the USSR, 71 for the United States, and 88 for China, which became in 1983 the major wheat producer of the world. The twelve-member EEC produced 83 million tons (France alone, 33) and India, 45 (World Wheat Council, Market Report, July 1, 1986).

plained by regional specialization within the United States. Fifty-five percent of wheat cultivation takes place in the Great Plains, 20 percent is located in the Northwest, and 15 percent is produced as a side crop in the Corn Belt. In all these states except the Corn Belt, rainfall is insufficient to cultivate corn and soybeans. Eighty percent of all wheat production (durum wheat excluded) consists of winter wheat, and of this, 60 percent is Hard Red Winter.

Over a twenty-year period, from 1960 to 1980, the harvested area of winter wheat increased by 27 percent, or about 4.4 million hectares, and the area planted with spring wheat increased by 51 percent, or about 2.1 million hectares (Table 6.2). The cultivation of spring wheat is practiced primarily along the Canadian border (in Canada, spring wheat still represents more than 90 percent of total wheat production).

Yields during this twenty-year period increased by 43 percent for winter wheat and by 64 percent for spring wheat. This last figure (2.6 percent per year), comparable to the yield increases experienced with corn (2.5 percent per year), is impressive for varieties that cannot be hybridized as corn can, a factor that has permitted an extraordinary expansion in the cultivation of corn. Its higher rate of yield growth explains why spring wheats represented 19 percent of American wheat production in 1980 as compared with 15 percent in 1960.

TABLE 6.2. U.S. wheat: Twenty-five years of progress

	Harvested Area (millions of hectares)	Yield (ton/ha)	Production (millions of tons)
Winter wheat			
Average, 1959–61	16.1	1.74	28.0
Average, 1974–76	19.7	2.12	41.8
Average, 1979–81	20.5	2.48	50.7
Average, 1983–85	19.9	2.69	53.4
Growth rate, 1960–75	+1.3%	+1.3%	+2.0%
Growth rate, 1975–84	+0.1%	+2.7%	+2.8%
Spring wheat			
Average, 1959–61	4.1	1.17	4.8
Average, 1974–76	5.9	1.73	10.2
Average, 1979–81	6.2	1.92	11.8
Average, 1983–85	5.0	2.27	11.3
Growth rate, 1960–75	+2.6%	+2.6%	+5.2%
Growth rate, 1975–84	−1.8%	+3.1%	+1.1%

Source: USDA, *Agricultural Statistics, 1977,* Table 3; USDA, *Agricultural Statistics, 1982,* Table 3; USDA, *Wheat Outlook,* September 11, 1985.

These results, both for winter wheat and spring wheat, become even more impressive when one considers that land area was increased at the same time, which normally leads to the cultivation of lower-quality land, in particular that placed in reserve until 1973 under programs to limit production.

Granted, the use of irrigation also increased, particularly on the West Coast, where yields exceed 5.0 tons per hectare, compared to 3.5 tons per hectare in nonirrigated areas; even in Kansas, the largest wheat-producing state, irrigated areas increased from 300,000 to 800,000 hectares between 1960 and 1975. Yet the vast majority of wheat production continues to take place in the form of dry-farming, that is, crop rotation very often including a fallow period every other year. Farms in Kansas, the largest winter wheat-producing state, measure on the average 268 hectares, and those in North Dakota, the largest spring wheat-producing state, average 406 hectares. By comparison, the average for the entire United States in 1979 was only 180 hectares. This considerable difference is explained largely by the need for bian-nual fallow periods.

Even more significant is the development during the 1970s of "soft" wheats, such as the Soft Red Winter, in the Corn Belt and in the Southeast. The share of this wheat in total production increased from 13 percent in 1970 to 26 percent in 1981, and its share of exports increased from 3.5 to 24 percent. But its share in the 1986 production decreased to 15 percent because of adverse economic and climatic conditions.

The increase in wheat production in the United States is more likely to come from areas with intensive production of "soft" wheat, where yields have increased from 3.5 to 6.0 tons per hectare (the Corn Belt, the Southeast, and the Northwest), as well as from double cropping of wheat and soybeans, which currently is taking place on only 15 percent of existing soybean lands and could be substantially developed in the Southeast.[3] This increase, however, will show up only if economic conditions are favorable.

The vast extension and high labor productivity of wheat cultivation in the United States can be illustrated by a farm in a northwestern state, where on an area of 320 hectares an average annual income of $20,000 was obtained in 1972 by a farmer using no outside labor, except for his two daughters, who each worked fifteen days a year during the harvest seasons. Furthermore, this farmer had important responsibilities with his association, spending 119 days in Washington, D.C., while continuing to operate his farm eight thousand kilometers away. Such cases are not rare.

The opening of the Soviet market in 1972 following the detente established between Nixon and Breznev signaled a great expansion in American wheat production affecting all categories. This was the "all-out" policy extolled by Republican agriculture secretary Earl Butz which from 1974 to 1977 lifted all planting restrictions in effect at that time.

The combined increases in land area and yields led the Carter administration in 1978 to reestablish a set-aside requirement of 20 percent of wheatland for all farmers wishing to participate in the price support program. This resulted in a reduction of 14.6 percent in planted land area and a reduction of only 12.2 percent in the 1977–78 harvest because yields increased by 3 percent. The reestablishment of this voluntary set-aside was renewed in 1979, although the initial stock, which reached 32 Mt on June 1, 1978, had diminished to around 25 Mt by June 1, 1979. This still represented half of annual production. Nonetheless, Secretary of Agriculture Robert Bergland let it be known in July of 1979 that there would be no further set-aside program in 1980. That meant that 5 million hectares were placed back under cultivation. Despite the embargo ordered in January 1980, limiting deliveries to the Soviet Union to 8 Mt, no set-aside was ordered for 1981.

The rapid production growth rate also led the Carter administration to create a program of stocking wheat reserves on the farm. Farmers could take out a loan at an advantageous rate for long-term stockpiling whenever the prices for the crop year 1980–81 fell below $154 per metric ton (140 percent of the support price), and as long as the maximum volume set by the secretary of agriculture for this "farmers-owned reserve" was not attained. They would have to destock when prices exceeded 175 percent of the support price, or $193 per metric ton. This program permitted the neutralization of 9.3 Mt of wheat on June 1, 1978, which increased to 11.2 Mt on June 1, 1979, and decreased to 6.8 Mt on June 1, 1980, since beginning in June 1979 world prices exceeded the trigger price for destocking. By 1981, however, the farmers-owned reserve began to increase again, reaching 13 Mt on June 1, 1982, and 27 Mt on June 1, 1983. Thus the PIK (payment in kind) Program, established by Republican

Secretary John Block, allowed a reduction of 13 million hectares of wheat land and 12 percent, or 11 Mt, of production (66 Mt for the PIK year 1983–84). Total stock levels for wheat were thus maintained below 42 Mt, that is, 60 percent of annual production, for the third consecutive year. Thanks to the allocation of 14 Mt of "public" wheat to farmers who agreed to receive in-kind compensation, the farmers-owned reserve was reduced to 18 Mt by June 1, 1985, a level below the ceiling of 19 Mt set by the secretary of agriculture.

According to the World Wheat Council, however, in spite of the set-aside program of 30 percent (10 percent unpaid diversion, 20 percent paid) equivalent to 8.5 M ha, production for 1984–85 has increased to about 71 Mt—in other words, exactly the average level reached in 1980–82 before PIK. The PIK program for wheat may end up having been nothing more than a short-lived remedy. A freeze on support prices was deemed necessary, and the interim law of April 4, 1984, inspired by Undersecretary of Agriculture Daniel Amstutz enacted it. The objective of the 1985 farm bill was to prolong this rigorous policy. Thanks to it, the 1985–86 production was reduced to 66 Mt, but this reduction was not sufficient to prevent an increase of 4 Mt in the U.S. wheat stocks, which on June 1, 1986, were again 9 Mt above the desired limit of 42 Mt (see Appendix 20).

The Corn Belt

American agriculture has gained world renown for its extraordinary Corn Belt, where more than 70 percent of U.S. corn production is concentrated, or one-third of world maize production. Corn is produced in rotation with soybeans in the following six states in descending order of production: Illinois, Iowa, Indiana, Nebraska, Ohio, and Minnesota.

Since 1976, American production has regularly exceeded 150 Mt, reaching 225 Mt in 1985–86, of which 30 to 55 Mt were exported. About 45 percent of world production and 70 percent of world exports are thus concentrated in the United States. The world's other producers are far behind, as shown in Table 6.3, in both production volume and, with the exception of the European Community and Yugoslavia, yields.

For twenty years, from 1960 to 1980, advances in corn production in the United States were considerable, as shown in Table 6.4, and they have continued since the world economic crisis of 1973 and 1981. These production advances are particularly interesting in that they are almost entirely attributable to growth in yields, which remained from 1960 to 1975 at 2.5 percent per year, double the percentage observed for winter wheat (1.3 percent per year). Furthermore, this annual growth rate appears to be increasing—the average annual crop yield from 1979 to 1981 amounted to about 6.6 tons per hectare, which corresponds to a growth rate of 4.8 percent per year. This is substantially greater than the rate for winter wheat (3.2 percent).

A forecast conducted by the University of Minnesota raises the hope that corn yields might increase by 0.08 tons per hectare per year until 1990, essentially because of classic plant selection (0.06 tons per hectare per year), irriga-

TABLE 6.3. World production of corn, 1979–1981

	Production (millions of tons)	Yield (ton/ha)	Harvested Area (millions of hectares)
United States	193	6.5	30
People's Republic of China	61	3.0	20
Brazil	19	1.7	11
EEC (France and Italy)	16	5.9	3
Mexico	12	1.7	7
Romania	12	3.6	3
South Africa	11	1.9	6
Yugoslavia	10	4.4	2
Argentina	10	3.2	3
USSR	9	2.8	3
World	421	3.3	130

Source: FAO, Production Yearbook, 1981, Table 13.

tion and parasite control (0.02 tons per hectare per year), then by 0.19 tons per hectare in the year 2000 because of biogenetics.[4]

The area planted with corn is subject to substantial fluctuations, however, either as a result of voluntary land set-aside programs (in 1968, 1972, 1978, and especially 1983) or of the price decrease (in 1961 after the decision by the Kennedy administration to set the support price at the world price level). Yields suffered the effects of corn blight in 1970 and droughts in 1964, 1974, and 1983. The standard deviation is thus equal to ±17 percent, which is considerable in terms of world market equilibrium because such a variation can represent between one-half and two-thirds of the quantity available for exports by the world (33 Mt out of 55 to 75 Mt).

These constraints explain the Carter administration's interest in a long-term reserve system and in an international agreement aiming to share the cost of financing stocks. Stocks had indeed increased considerably in recent years, rising from 10 Mt on October 1, 1976, to 22.5 Mt, 28 Mt, 32.7 Mt, and 42.8 Mt respectively in the following four years. Under the Reagan administration, stocks continued to increase from 32.0 Mt on October 1, 1981, to 58.1 Mt on October 1, 1982, and to 79.3 Mt one year later.

Thus, as in the case of wheat, but with less success for the set-aside of land and with greater success in long-term stockpiling on the farms, programs were established in 1978 and renewed in 1979, 1983, and 1984 to cut production. In 1977–78 the set-aside program was announced too late. It required all producers wishing to benefit from the price guarantees to set aside 10 percent of corn land in addition to an optional supplementary reduction of 10 percent in exchange for a payment. In reality, the area decreased by only 5 percent (or

TABLE 6.4. U.S. advances in corn production

	Harvested Area (millions of hectares)	Yield (ton/ha)	Production (millions of metric tons)
Average, 1959–61	26.8	3.60	95.9
Average, 1974–76	27.2	5.19	142.2
Annual rate of increase over fifteen years (%)	0.1	2.5	2.7
Average, 1979–81	29.3	6.57	192.9
Annual rate of increase over five years (%)	1.5	4.8	6.3

Source: USDA, *Agricultural Statistics, 1982*, Table 32.

1.7 million hectares) since only 60 percent of corn producers participated. Farmers who refused to sign were not afraid of forfeiting a possible payment because the market price was not expected to drop below the target price level ($83 per metric ton) that triggered deficiency payments.

Based on projections of another record harvest in 1979, President Carter himself announced in November 1978 a set-aside program identical to that of the preceding year. The rate of participation was even lower than that in 1978; because prices rebounded sharply with the announcement of a poor Soviet grain harvest, participation amounted to a mere 20 percent. No set-asides were ordered in 1980 for feed grains, which permitted the reactivation of 2.5 million hectares of cropland.

On the other hand, loans for long-term stocks (three to five years) made available as soon as prices fall below $110 per Mt (125 percent of the loan rate), to be reimbursed as soon as prices exceed $128 per Mt (145 percent of the support price), were highly successful. On October 1, 1978, the farmers-owned reserve reached 8 Mt (prices reached $63 per metric ton that month); on October 1, 1979, it reached 13.7 Mt (prices had climbed back up to $77 per metric ton); on October 1, 1980, it came to 25.4 Mt, thus representing 60 percent of total U.S. corn stocks evaluated at 42.8 Mt, including the 7.6 Mt bought by the CCC partly as a result of the embargo on the Soviet Union.

The drop in production in 1980–81 (−15 Mt in the United States, −10 Mt for the entire world) resulted once again in a more modest total ending stock for October 1, 1981 (32 Mt). The record harvests of 1981 and 1982 (208 and 212 Mt), however, and the decrease in exports (50 Mt compared to an average of 60 Mt in 1975–78) led to stocks of 58.1 Mt and 79.3 Mt. It was then that the Reagan administration decided to establish the PIK program. The 1983 harvest was reduced to 106 Mt, and the stocks as of October 1, 1984, fell to 18.1 Mt. To compensate for the nonproduction of 18 million hectares voluntarily laid fallow, corn producers were paid in kind. They received 45 Mt of corn taken from public stocks or from the farmers-owned reserve.

Although the PIK program for wheat was not very effective, the one for corn brought lasting improvement to the world market for feed grains. Despite the 1984 harvest, which exceeded 194 Mt, stocks as of October 1, 1985 were still limited to 41.9 Mt; but even though 3.2 M ha were set aside, the 225 Mt record harvest of 1985 put the closing stocks at 101.2 Mt on October 1, 1986, that is, above the level of the year preceding the PIK program.

These impressive results have placed the United States in a position of world leadership in the production of corn. They were not achieved, however, without a considerable public and above all private research effort, recently recognized when corn geneticist Barbara McClintock received the Nobel Prize for medicine in 1983 at the age of eighty-one. After the discovery of the first methods of corn hybridization in 1920, private companies, beginning in 1945, have done most of the work in research, development, and extension. The use of hybrid corn throughout the world is mostly a product of the work conducted by these companies between 1945 and 1955.

The names of the large American firms involved in this selection work are well known: De Kalb and Pioneer produce about 55 percent of the hybrid seeds; Cargill, Funk, Ferry-Morse, Trojan, Acco, and Northrup King share the rest of the market. Following the enormous profits realized between 1972 and 1976, pharmaceutical companies such as Pfizer and Ciba-Geigy have attempted to acquire selection firms.[5]

An example will illustrate the quick response of corn selectors. Hybridization was made economical through the discovery of male sterility varieties, in which the male flower (which differs from the female corn flower, as is not the case of wheat or soybeans) is degenerate and infertile. It is therefore possible to sow in the same field two rows of variety A with a fertile male flower alternated with six rows of variety B with a sterile male flower and a fertile female flower. The product "A male–B female" will be found on all the B rows; it is the desired hybrid.

During the corn blight of 1970,[6] it was observed that the varieties resistant to this disease could not benefit simultaneously from the quality of male sterility. It was necessary to return to the old methods of manual castration. Thus, within a few months, trailers with dozens of elevated seats on the right and left of the tractor were manufactured and thousands of students were hired to do the work of castrating male flowers, which fortunately are situated at the top of the corn stem; thus more than 50 percent of the sowing of the following year was made possible by hybrids resistant to corn blight.

This example shows the strength and responsiveness of the American corn production system, which, after rivaling sugarcane and beets in the production of corn-fructose and petroleum-producing countries in the production of alcohol, is now turning to biotechnology, using leafy corn varieties developed by De Kalb–Pfizer Genetics. Pioneer Hi-Bred International is branching out in the selection of very short-cycle hybrids (seventy-five days), which would allow the extension of corn production in Canada.

The Soybean Miracle

When Clarence Palmby was assistant secretary of agriculture in 1971, he frequently spoke of the "soybean miracle" to illustrate the responsiveness of this product to market conditions. Starting in 1969, when the support price of soybeans was reduced by 10 percent, the level at which it remained until 1976, soybean exports and production have continually increased. Soybeans became a pet subject for the Nixon administration because this product always responded according to free market economic theories, sacred within the Republican party and the powerful farm association, the Farm Bureau.

Soybeans were known as early as 2838 B.C. in China and were discussed as early as A.D. 1740 in Paris. It was not until 1924, however, that their cultivation in the United States reached 200,000 hectares. Soybeans did not become economically significant until 1952, with a harvest of 6 million hectares, when the production of soymeal for livestock became the driving force in soybean development. Until that time, germinated soybeans were used in the preparation of protein-rich salads, highly appreciated in the Far East, or as a source of light, tasteless oil, popular in the United States. The residue from oil pressing, the oil cake, became the principal product beginning in 1952.

As in the case of corn, the dominant position held by the United States in soybeans is overwhelming. Despite competition from Brazil, American production, of which nearly 60 percent is exported (either in the form of meal or of beans), represents nearly 60 percent of world production, far ahead of all other producers (Tables 6.5 and 6.6).

In recent years, however, Brazilian production (1972–77 average: 9.2 Mt; 1978–83 average: 16.6 Mt) has become a significant alternative for purchasers of oil cakes. Indeed, this country has developed a policy based on what economists call the "scissors effect." A tax is placed on exports of Brazilian soybeans, and the revenues from this tax are used to finance exports of soymeal. This policy allows Brazilian plants, most of which are owned by the large international shippers, to operate at near capacity and permits the Brazilian population to benefit from all soy oil produced. Importing countries therefore have an incentive to buy beans for their own crushing plants from the United States but to buy meal for their livestock feeding operations from Brazil. Such is the case for France, the world's largest importer of meal (2.2 Mt of soymeal in 1979–80 compared to only 0.8 Mt of beans).

The growth in Argentine production (1972–77 average: 0.7 Mt; 1978–83 average: 3.7 Mt) has been rapid. The 1983–84 harvest reached the record level of 4.7 Mt and 1984–85 a new record with 6 Mt. Exports exceeded the equivalent of 3 Mt of soybeans in 1980–82.

In general, soybeans require a warm, humid climate. For this reason they would be difficult to cultivate in Europe or North Africa without resorting to costly irrigation. The countries of Asia do not appear up to now to have considered soybeans as an export crop or a means to develop their livestock production.

TABLE 6.5. The world's principal producers of soybeans

Country	Year	Production (millions of tons)	Yield (ton/ha)	Year	Exportation Beans (millions of tons)	Exportation Meal (millions of tons)	Exportation Bean Equivalent (millions of tons)[a]	Ratio Export: Production (%)
1. United States								
	Average, 1972–77	37.4	1.81	1975–76	15.1	4.7	21.0	56.2
	Average, 1978–83	55.6	2.00	1979–80	22.5	6.7	30.9	50.1
	1983–84	41.8	1.67	Ave. 1980–82	23.1	6.3	31.1	57.0
2. Brazil								
	Average, 1972–77	9.2	1.65	1975–76	3.4	4.2	8.7	94.4
	Average, 1978–83	13.6	1.61	1979–80	1.5	7.5	10.9	70.1
	1983–84	15.3	1.73	Ave. 1980–82	1.2	8.3	11.6	81.0
3. China								
	Average, 1972–77	9.4	1.05					
	Average, 1978–83	8.8	1.09					
	1983–84	9.5	1.17					

4. Argentina							
Average, 1972–77	0.7	1.72	1975–76	0.1	0.3	0.5	68.0
Average, 1978–83	3.7	1.93	1979–80	2.8	0.3	3.2	83.6
1983–84	4.7	1.92	Ave. 1980–82	2.4	0.5	3.0	81.0
5. Paraguay							
Average, 1972–77	0.2	1.55					
Average, 1978–83	0.6	1.52					
1983–84	0.6	1.49					
Others							
Average, 1972–77	3.3	1.01					
Average, 1978–83	4.9	1.13					
1983–84	5.6	1.15					
World							
Average, 1972–77	60.2	1.54	1975–76	19.3	11.3	33.5	55.7
Average, 1978–83	81.2	1.73	1979–80	28.0	18.0	50.6	53.1
1983–84	77.5	1.56	Ave. 1980–82	27.6	20.7	53.7	62.0

[a]Bean equivalent tonnage = bean tonnage + 1.26 × meal tonnage.
Source: "World Oilseed Situation and Outlook," *USDA Foreign Agriculture Circular*, September 1983.

TABLE 6.6 Exports of the three major soy export countries (millions of tons)

	Average, 1975–77[a]		Average, 1978–80[a]		Average, 1981–82[a]	
	Millions of tons	Percentage	Millions of tons	Percentage	Millions of tons	Percentage
Soybeans						
United States	16.7	83	21.3	85	24.9	90
Brazil	2.5	12	1.2	5	1.0	4
Argentina	0.9	4	2.6	10	1.6	6
Soy meal						
United States	4.8	48	6.3	48	6.3	41
Brazil	5.0	50	6.5	49	8.1	52
Argentina	0.3	2	0.4	3	1.1	7
Total in soy meal equivalent[b]						
United States	18.1	69	23.2	70	26.1	70
Brazil	7.0	27	7.5	23	8.9	24
Argentina	1.0	4	2.5	7	2.3	6
Total three countries	26.1	100	33.1	100	37.3	100

Sources: "Oilseeds and Products," USDA *Foreign Agriculture Circular*, March 1979, May 1980, September 1983.
[a]Year from October 1 to September 30.
[b]Soy meal equivalent: soy meal equivalent tonnage = soybean tonnage × 0.795 + soy meal tonnage.

American production expanded initially in the Corn Belt, where soybeans are cultivated in rotation with corn: one-third soybeans, two-thirds corn, this proportion varying depending on respective price levels and the timing of corn planting. For example, if in April rainfall is too low to plant corn, soybeans can still be planted one month later.

The price of soybeans must be attractive, that is, equal to that of corn multiplied by a coefficient varying between 2.2 and 3.0 according to the region and climatic conditions. For the United States as a whole, average yields for the last seventeen years have been 5.57 tons per hectare for corn and 1.86 tons per hectare for soybeans, or a ratio of 3 to 1.

On a theoretical level, an agronomist from the University of Illinois, Bruce Vasilas, has calculated that the maximum yield of a corn hybrid could not exceed 30.7 tons per hectare, whereas that of soybeans would be 15.1—a ratio of 2 to 1. Record average yields for Illinois are 8.4 and 2.6 tons per hectare respectively—a ratio of 3.23 to 1; the absolute individual records are 21.2 and 5.6 tons per hectare—a ratio of 3.78 to 1.

Many farmers in the Corn Belt closely follow prices on the Chicago futures market for their coming harvests. They do not intend to speculate on the prices of corn and soybeans, although some may be interested in hedging, selling futures for a part of their harvest to obtain loans for the crop year from their bankers at attractive interest rates. According to the Commodity Futures Trading Commission, only 5.6 percent of farmers bought or sold futures contracts in 1973. But they all turn to Chicago to know if they should emphasize soybeans by planting them as a full third of their crop, or if they should limit soybean crop area because of Brazilian competition and increase production of corn, gambling on a poor Soviet grain harvest. For this reason, Midwest radio stations broadcast the last available prices on the Chicago Board of Trade at all hours of the day.

Before they faced competition from the southern states of Brazil (the states of Rio Grande do Sol, Parana, Mato Grosso, and Goias), soybean farmers in the Corn Belt had to compete with cotton farmers in the Mississippi Delta. These farmers had discovered soybeans in 1966, when the price of cotton fell by more than 30 percent and cotton land area diminished considerably.

Thus, in 1978, a fairly representative year in which production was high and prices moderate, soybean production was distributed as follows: 57 percent in the Corn Belt (such as Illinois and Iowa), 31 percent in the Mississippi Delta (such as Missouri and Arkansas), and 5 percent in the Southeast (North and South Carolina and Virginia). In 1983, the Corn Belt produced only 44 percent of all soybeans. Development of double cropping in wheat and soybeans in the Southeast explains this change. More than half of American soybean production now takes place outside of the Corn Belt. This extension of soybean cultivation toward the South and East has not diminished Corn Belt production but simply supplemented it. The same occurred when Brazil and later Argentina entered the world market.

One of the most striking aspects of the miraculous use of proteins in livestock feeding is that market growth has remained ahead of production. This is confirmed by the level of American soybean stocks on September 1 of each year. Approximately every six years, these levels decline regularly to below 10 percent of American production, causing periodic tension on prices. This was the case in 1963–67 (average stock equaling 5 percent of production), again in 1971–73 (6 percent), and once again in 1977–79 (9 percent). From 1980 to 1983 there was a return to lower price tension, with stocks equaling 18 percent of production, a figure close to the average for 1974–76 (14 percent) and for 1968–70 (23 percent). On September 1, 1983, stocks amounted to 25 percent of production, and three years later to 27 percent, and the prices were down to the loan rate level ($185 per ton).

There is another reason for the expansion in world soymeal demand. Many countries have adopted only the grain production and consumption side of the American model; they have developed intensive husbandry of poultry or cattle, but their concentrated feed industry is too embryonic to enable them to meet the demand for livestock feed. Thus there is a considerable potential in these countries for the import of vegetable proteins. Appendix 16 provides a list of the world's sixteen largest soy user markets. Besides Western Europe and Japan, increasing demand can be seen from USSR, Eastern Europe, and Southeast Asia, following the pattern established by Japan.

Given the rapid increase in demand, one can imagine the tension that would have resulted on the world market if the supplies from the Mississippi Delta, Brazil, and Argentina had not been added to supplement Corn Belt production. There is clearly room for additional producers of vegetable proteins in the world.

Another reason why newcomers are necessary is that for soybeans, unlike corn and wheat, increased cropland area was a much more important factor than yield increases in the U.S. production growth observed between 1960 and 1975 (Table 6.7) and even from 1975 to 1980, when yields increased substantially.

The problems in improving soybean yields are characteristic of this leguminous plant, for which hybridization is currently difficult because fertilization requires a manual operation. Great advances were made during the 1950s in perfecting inoculums such as rhizobium, which allows the assimilation of nitrogen contained in the soil and air. This is another miraculous effect of soybeans: instead of requiring nitrogen fertilizers, soybeans enrich the soil in nitrous substances, making them a highly desirable crop for rotation with corn or cotton.

Since 1971, when male sterility varieties were discovered, the fertilization of soybean seeds has been facilitated by the introduction of hives of a species of small bees, which, when placed near a selected field, allow the creation of higher-yield, more disease-resistant varieties with maximum protein content and minimum oil content. Another paradox of this miracle plant, even though soybeans yield less oil than any other oilseed, in 1977–78 they provided 41 percent of the world's production of edible oil and 22 percent of edible fats.[7]

Finally, methods of harvesting soybeans have improved substantially in recent years with the perfection of cutting blades better adapted to the crop,

TABLE 6.7. The development of soybeans in the United States, 1945–1983

Evolution of Soybeans from 1945 to 1980s	Harvested Area (millions of hectares)	Yield (ton/ha)	Production (millions of tons)
Average, 1944–46	4.1	1.299	5.3
Average, 1959–61	9.8	1.635	16.0
Average, 1974–76	20.8	1.754	36.5
Average, 1979–81	27.7	1.990	55.2
1982	28.3	2.190	61.9
1983	24.9	2.020	50.1
Annual rate of increase (%)			
1945–60	5.98	1.55	7.64
1960–75	5.13	0.47	5.65
1975–80	5.89	2.60	8.63

Source: USDA, Agricultural Statistics, 1983, Table 178.

which has branches very close to the ground. Before this advance, nearly 15 percent of the harvest was lost by using corn pickers, which did not cut close enough to the ground.

Since 1973, intensive research efforts have been undertaken by universities, the USDA, and private firms. In 1976 there were thirty government researchers and thirty private researchers specializing in soybean selection, whereas in 1970, there was only one.[8] Yields increased from 1975 to 1980 at an annual rate of 2.6 percent, a pace as rapid as that for spring wheat, whereas from 1960 to 1975 soybean yields increased five times slower than those for spring wheat.

New derivatives of soybeans are taking on an economic importance unknown in the 1960s. Four degrees of vegetable protein concentration are now offered on the American market and, more recently, on the European market: (1) soymeal with a 44 to 50 percent protein content, the classic livestock feed; (2) soy flour with less than 65 percent protein; this is an adjuvant in products based on grain, in baking and confection, for which it is used as an emulsifier, a foam, and an antioxidant, or as formula for babies who cannot tolerate cow's milk; (3) soy protein concentrate containing 65 to 90 percent protein, which is produced in an extruded or textured form and supposedly is intended to rival the fiber in meat; known as texturized vegetable proteins (TVP), these concentrates have met with some success in the United States, although less than some expected in 1975; and (4) isolated soy protein with a protein content greater than 90 percent, which is intended to be mixed with animal protein so as to lower its price; sausages and ground meats containing from 3 to 30 percent of this material may thus be introduced on the food market.

These new products will not shift the major use of soybeans to human consumption. The soymeal, as a means of nourishing livestock with protein, remains the most competitive use of this crop, and it is in this form that world soybean production is increasingly being used.

The 1973 Soy Embargo

Clarence Palmby was no longer assistant secretary of agriculture on June 27, 1973, when President Nixon ordered an embargo on all soy exports from the United States. This extraordinary political blunder is difficult to explain. As we have seen, soybeans are even more miraculous from an economic standpoint than from a biological one. As Watergate began to loom on the horizon, soybean prices in the United States skyrocketed, in part because of the disappearance of anchovies, an animal feed protein raw material competing with soy protein, off the coast of Peru. Within a few weeks, soybean prices increased from $110 per metric ton to $440 per metric ton. This phenomenon preceded by a few months the petroleum crisis, which would also lead to an embargo and a quadrupling of prices.

In 1973, the American government had no means of control over or even information on agricultural exports. Since that date, a program of export monitoring has been established requiring American traders to declare all large

FIGURE 6.1 Price paid to farmers for a metric ton of soybeans, 1960–1980

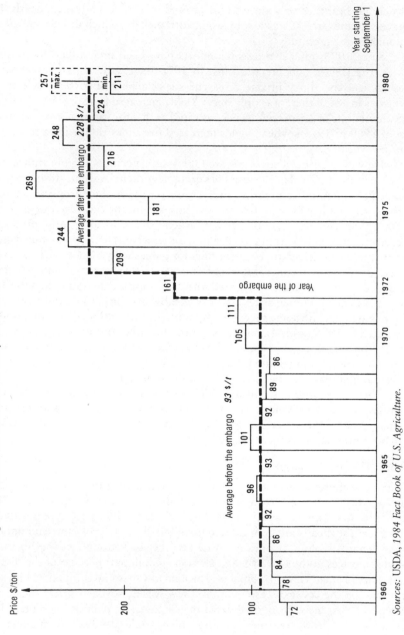

Sources: USDA, 1984 Fact Book of U.S. Agriculture.

contracts.[9] President Nixon, politically unstable and harassed by a burgeoning inflation rate, ordered an embargo, which brought immediate reproaches from the government of all client countries and caused a lasting loss of confidence in the capacity of the United States to remain a reliable supplier to importing countries. The embargo lasted only two months, and the tonnage of American soybean and soy meal exports was not much lower the year of the embargo than other years. Yet world confidence had been undermined.

With respect to prices, however, the embargo had a highly favorable result for American producers. The quadrupling of prices did not last, but prices did remain at about twice the previous level, confirming the place of soybeans as the number one cash crop for American farmers and the prime source of their monetary income (about $13 billion) (see Figure 6.1).

Sunflowers and Tobacco

American farmers are now turning to a new oilseed: sunflowers, perhaps a by-product of Soviet-American or Franco-American cooperation. Production is based on hybrid seeds, which, as in the case of corn, involve use of the sterile male cytoplasm discovered by INRA (the French national institute for agronomic research), thereby facilitating seed production.

In 1979–80 American yields (1.51 tons per hectare) were considerably greater than those in the Soviet Union (1.24) and Argentina (0.94). The United States has been the world's second largest producer with 3.5 Mt, behind the Soviet Union (5.4 Mt) but well ahead of Argentina (1.7 Mt). Cropland devoted to sunflowers doubled in the United States each year from 1975 to 1980, while it remained the same in the Soviet Union and increased by only 8 percent in Argentina. By 1980–81, however, a saturation point had been reached. Cropland devoted to sunflowers was reduced by 20 percent and production was down to 2.7 Mt. Since 1984, however, there has been a new type of double cropping in the northern portion of the Corn Belt: sunflowers are planted in early July after the harvest of winter wheat, with production of 1.0 ton per hectare on the average. Nevertheless, 1984 production reached less than 1.9 Mt and the average yield 1.16 tons per hectare.

There has also been much discussion of the high protein content of tobacco, although the economic future of this crop has been disputed. Planted directly and in high density (375,000 plants per hectare instead of 20,000), tobacco could be cut every six weeks, with a harvest of 165 metric tons of fresh leaves per hectare, according to S. G. Wildam of the University of California. He foresees yields of at least 3 metric tons of pure protein per hectare, or three times that obtained for soybeans, without considering the by-products.[10]

The Role of Beef Producers and Feedlots

No description of the technological pillars of the American model would be complete without a glance at the intensive method of cattle raising, the feedlot. This system has not yet been widely used in Europe, unlike the American pork and poultry raising models, which have been so well assimilated in Europe for a number of years that, for now, anybody could doubt whether the best models are not in Europe.

In 1980, beef producers received $2 per kg of carcass in Europe, $1 per kg in North America, and $.50 per kg in Australia. But these approximate prices are the result of totally different husbandry methods and meat quality.

Beef in Europe is essentially a by-product of dairy production because the cattle population is primarily composed of milk breeds. In Australia, it is based on extensive livestock husbandry; herds of breeds specialized for meat production rely on pastures for their food, and the number of head varies according to the amount of rainfall for each year.

The American system is about 75 percent based on the use of feedlots for livestock finishing. Unfatted stock generally purchased from western ranches arrive in the feedlots either at the age of seven months, just after weaning, or at the age of one year, weighing 350 kg in both cases. They are then placed in homogeneous groups of two hundred to four hundred head in large pens which are constantly supplied with fodder, grain, and soymeal. Live weight gains are about 1 kg per day, and young cattle are taken to the slaughterhouse by group as soon as they have reached a weight of 500 kg, at about fourteen to eighteen months of age. The desired result is classification as "prime," which corresponds to the top third in quality.

On January 1, 1978, there were 116 million cattle in the United States (132 in 1975, 111 in 1979, and 116 again in 1982) of which 37 million were Hereford and Angus meat-breeding cows and 11 million were Holstein dairy cows; these cows gave birth to 44 million calves in 1978; 1,250,000 calves were imported (700,000 from Mexico, 525,000 from Canada) and 120,000 were exported, principally to Canada. Forty million head were slaughtered, of which 8.5 million were culled cows, 27.7 million were young cattle from the feedlots, and 2.5 million were young cattle raised by the extensive method. Cattle raised in lots thus represented 70 percent of slaughtered livestock.

There were 127,000 feedlots in 1978 (157,000 in 1975) out of which 125,000, each accommodating less than 1,000 head, produced 32 percent of the total production of feedlots (these small operations are located primarily in the Midwest on grain farms). The 415 largest units, ranging from 8,000 to more than 100,000 head, accounted for 52 percent of the total production from feedlots (these large specialized operations are located in the arid portions of Texas, Colorado, California, Nebraska, and similar states).

In 1975 it was thought that the optimal number of cattle per feedlot was about 12,000 head per year, which is intensive husbandry, allowing about 6,000 head to be fatted twice a year, but this notion has altered with time. The specificity of this approach, and its disparity from that used in Europe, invites reflection on its origin. In reality, the situation is fairly clear and can be easily analyzed in economic terms:

1. Demand for one specific type of meat is fairly inelastic in relation to its

price, but comparative demand elasticity between different types of meat is high.

2. The meat producer is caught in a crossfire between the demand for meat and that for grain and soybeans, which have elasticities.[11] The great variability in the price of inputs has been accentuated since the beginning of the 1970s: corn blight in 1970–71 (causing a 15 percent corn price increase) was followed by a record harvest in 1971–72 (forcing prices from $1.33 to $1.08 a bushel of corn) and by a sudden increase in feed grains exports in 1972 (from 27.1 Mt in 1971–72 to 43.1 Mt in 1972–73).

3. The biological cycle is complex. In the short term, supply is increased by more intensive feeding, leading to greater weight at the time of slaughter. In the long term, the herd must be increased: in this case, four years can pass before the growth phase is stabilized, and production cycles will thus persist for about eight to ten years for beef (four years only for pork).

These three characteristics of cattle production in the United States result in high sensitivity to price variations for meat producers.

An analysis of production cycles shows that the price of beef varies considerably from one quarter to the next (± 10 percent of the annual average) and from one year to the next (± 20 percent of the average over five years). These price variations influence the number of animals raised in feedlots; this number decreased from 28 million in 1972 to 21 million in 1975, climbed again to 28 million in 1978, only to decline to 24 million in 1980. During off-peak years, feedlots are therefore partially empty for part of the year.

The Role of Adjustment by the Pork Producers

Such a system might lead to erratic supplies of meat to the population. But the American consumer is highly adaptable and falls back on pork or poultry when beef prices are high. Thus in 1979, beef production decreased by 6 percent and prices increased by 12 percent. Simultaneously, the production of pork increased by 10 percent with a 15 percent decrease in price. The consumption of poultry (chickens and turkeys) increased by 6 percent. Total per capita consumption changed only slightly, but its composition was substantially altered. Figure 6.2 shows the extreme flexibility of American consumption. Nonetheless, beef makes up 45 percent of total meat consumption—60 percent of red meat consumption—which makes beef production and prices the key element in all meat production, since pork and poultry production are easier to adjust.

There has been an extremely rapid evolution in the system of pork production in the United States. Two-thirds of American production comes from the Corn Belt, particularly Iowa and Illinois, which together contribute one-third of the total. Almost half a million farms raise hogs (9.5 million sows, 77 million hogs slaughtered in 1978), but in 1979, 40 percent of production came from units delivering more than a thousand head per year (in 1964, only 7 percent came from these large units, and in 1974 only 25 percent). In 1978, more than 40 percent came from operations using completely enclosed buildings.[12] Re-

FIGURE 6.2 **Per capita consumption of meat according to origin, 1965–1979 (in pounds)**

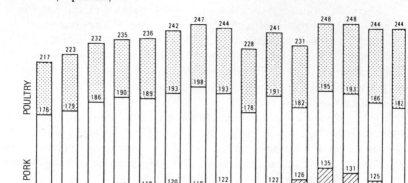

Source: Agricultural Statistics, 1982, Tables 473 and 715; *1979 Handbook of Agricultural Charts,* 198 and 215.

duced flexibility can thus be foreseen for pork production. Hog farmers will no longer be able to decrease or increase their capacity according to grain and beef prices.

The success of the productions basic to the American model have just been briefly examined: wheat, corn, soybeans, and intensive animal husbandry using prepared, balanced feed. The unifying element among all of these apparently unrelated products is the market price. The constant variation in prices has been the fundamental element in the technological advances made by American farmers, subjecting them constantly to the pressures of competition.

This system of domestic grain and soybean prices, in direct relation with world prices, and of meat prices, a function of the price of mixed feeds, has led to both specialization and flexibility. The results are highly productive farms, capable of taking full advantage of research and exporting not only their products but also their technologies.

This American model owes its existence and form to an economic and agronomic research policy established at the end of the last century by federal and state governments.

RESEARCH AND INNOVATION

In 1977, during renewal of the farm bill, a study on agricultural research was prepared at the request of the Senate Agriculture Committee. It revealed a broad consensus among the experts attributing the low proportion of income spent by Americans on food (15 percent, not including alcoholic beverages) to the success of agricultural research.

One hundred years ago an American farmer could feed, in addition to himself, five other Americans. In 1977, he could feed fifty-five other persons in the United States and throughout the world. The prospect of a world population of 6 billion by the year 2000 (4.5 billion in the Third World) confirmed the need for increased spending on public agricultural research.[13]

This praise of American agriculture, however, dissimulated a certain concern reflected by the September 29, 1977, act, which, once again, reorganized U.S. agricultural research, extension, and education.[14] In fact, U.S. agronomic success has been based on three elements unique to the American system: the network of fifty-three American agricultural universities, the Economic Research Service, and the network of private foundations.

The Land Grant Colleges

In 1862, President Abraham Lincoln, after creating the USDA, signed the Morrill Act. This law granted large amounts of federal land to any state creating a university that included an agricultural research center. The famous land grant colleges and their university campuses were thus born. The Hatch Act of 1887 provided funds to finance research programs.

The agricultural universities that emerged became famous worldwide for their contributions to agriculture, economics, and other fields. Some of the more famous institutions are located in Ames, Iowa; Ithaca, New York (Cornell); Champaign-Urbana, Illinois; Davis, California; Lincoln, Nebraska; Minneapolis–St. Paul, Minnesota; and Lafayette, Indiana (Purdue).

Today, more than six thousand research professors work in these fifty-three agricultural universities, which in turn are linked to more than seventeen thousand Extension Service agents, responsible for agricultural development.

The Federal Research Service of the USDA has three thousand researchers in two hundred laboratories organized in four large regional offices located in Beltsville, Maryland; Peoria, Illinois; New Orleans, Louisiana; and Berkeley, California. Private firms employ more agronomic researchers than the federal and state governments combined—more than sixteen thousand mid- and high-level employees.

This research effort is not exceptionally large for a nation with so extensive an agricultural sector as the United States. It should be recalled that in France, the National Agricultural Research Institute, INRA, employs more than 1,500 researchers, the technical institutes of the CEMAGREF and the ACTA employ nearly 1,000 researchers, and the tropical research divisions employ 700 researchers at the GERDAT-CIRAD and the ORSTOM, or about 5 percent of the total number of public and private researchers (67,000). In the United States, public and private agricultural research employs only about 4 percent of the total number of American researchers (25,000 out of 570,000 in 1975). With respect to federal funds alone, this imbalance is even more striking. During the years from 1970 to 1975, agriculture represented only 2 percent of the total federal research budget ($16 billion) whereas that for NASA and the Pentagon accounted for 70 percent of the budget and the Department of Health, Education and Welfare, 10 percent. In 1940, by comparison, agriculture received 40 percent of the total federal research budget (only $74 million at that time). It was as if the success of American agriculture allowed the government to disregard this sector.

One might assume that the private sector took up the slack, but this does not appear to be the case. The agricultural industry in the United States does not undertake an extraordinary amount of research. With the exception of the farm machinery sector, which spends 2.5 percent of sales on research, and the veterinary pharmaceutical industry, which spends 6.3 percent, the remainder of the agrofood industry (selection and vegetable, animal and human nutrition, pesticides and fertilizers, food technology) spends less than 1 percent of total sales on research. By comparison, the pharmaceutical industry (human) devotes more than 10 percent of its sales to research.

These trends seem to be reversing, however, because the private agrofood sector is considering an increase in research personnel from sixteen thousand to twenty-two thousand from 1980 to 1990. No increases in the number of public sector researchers are planned despite the recommendations in 1977 of the National Academy of Sciences (NAS) in favor of a 10 percent annual increase in the nutrition and food research budget.[15]

The only real innovation from the public sector in 1978 and 1979, in accordance with the NAS recommendations, was the creation of special grants for public or private teams in three promising areas: chlorophyll photosynthesis, nitrogen fixing by the roots of plants other than legumes, and genetic manipulation of plant cells to produce biological mutations. To these fundamental subjects, the Carter administration intended to add longer-range weather forecasts, improvement in the marketing of food products, and development of the role of cooperatives and family farms, even small ones.

Yet funding for all of these areas amounted to only $60 million over the three years 1977–80; NAS was recommending $50 to $75 million per year. In December 1984, Paul F. Knowles, a former professor at the University of California and chairman of the Scientific and Technological Council for Agriculture, emphasized the carelessness of American agricultural research—only nine thousand vegetal species are studied out of three hundred thousand and a greater diversification of farm production is needed.

The success of the American agricultural sector, therefore, can be attributed more to its exemplary continuity for over a century than to its annual access to government funding. The United States is the only country in the world to have so consistently accumulated such a vast amount of knowledge and above all to have transmitted it so rapidly to its farmers and food industries. Its success is also due to the professionalism of American researchers, who display an exceptional combination of training and ability to identify objectives and make available financial, material, and technological resources.

Modern agriculture is an uninterrupted sequence of states of imbalance because changes in technology and raw materials are occurring more and more rapidly. To operate successfully under these conditions requires a high level of training. The availability of professors in contact with the realities of agriculture is therefore indispensable.[16]

The concentration of nearly all agricultural training and extension and of half of all public agricultural research within a single, decentralized institution, the land grant colleges, has produced tremendous results. This example of integration might well be considered by many agricultural countries not fortunate enough to possess such a structure since 1862.

The Economic Research Service

Another reason for the success of American agriculture is the close integration of economic and agricultural policy concerns with technological research and extension efforts. The key participant in this policy is the famous USDA Economic Research Service, which employs 425 economists, three-fourths of whom are located in Washington in a building near the office of the secretary and his six assistant secretaries.

The Hatch Act of 1887 required all agricultural research laboratories receiving federal funds to disseminate all their results, both technical and economic, immediately and to send them free of charge to all potential users (farmers, cooperatives, associations, specialized magazines, libraries, and other experiment stations). The USDA paid over $1 million in 1978 to the Postal Service for this free dissemination of information mandated by the law.

The ERS[17] also perfected a system of providing information monthly on the condition of harvests, product by product, complemented by data disseminated by the Agricultural Statistical Service and by the seventy agricultural attachés of the Foreign Agricultural Service. During years of great agricultural instability, beginning in 1970, American harvest outlooks took on increasing importance. Outlook Boards,[18] committees composed of production specialists from the various offices of the USDA (ERS, Statistical Service, International Division, Extension Service), meet periodically in Washington, on a date communicated in advance to agricultural journalists; at 3:30 P.M. (about one hour after the close of the Chicago Exchanges) they announce their projections, which will be modified monthly following the same procedure. The system of the Crop Reporting Boards was criticized when unacceptable discrepancies between the projected stocks and the field inquiries concerning the level of soybean production arose in January 1984. Since 1980, corrections of about 2 percent, which is the level of the target sampling error, had accumulated, leading to a final stock overvalued by 2.5 Mt. A task group formed as a result of these incidents recommended in July 1984 that from then on significant differences between the field inquiries and the economists' calculations be published, taking Landsat satellite observations into account.[19]

Every year since 1923, a national agricultural projection conference, called the Agricultural Outlook Conference, has been held in Washington. Its aim is more to provide economic extension than to allow consultation between the various professional groups in attendance: agricultural organizations, universities, industrialists, distributors, bankers, brokers, journalists, and American and foreign government officials.

Although the ERS has thus revealed factors affecting the short-term economic situation, it has also conducted longer-term studies, some of which are specialized in economic projections for certain products. For example, Purdue University in Lafayette, Indiana, where President Nixon's second secretary of agriculture, Earl Butz, was a professor, specializes in the economics of corn, and the University of Minnesota, where President Carter's secretary, Robert Bergland, taught, specializes in dairy economics.

Based in the land grant colleges, the agricultural economics departments are in close contact with farmers through the Agricultural Extension Service, for which they carry out many responsibilities; they also work closely with food and agricultural firms that provide scholarships for students preparing doctoral theses. These contacts enable them to be more realistic and often allow them a more political approach to economic problems because of their financial independence and the varied political orientations of the states on which they depend.

The ERS also acts as a clearinghouse for the wide-ranging sources of economic information which have greatly helped integrate American agriculture, in both domestic and international markets. Therefore ERS publishes the least contested middle- and long-term studies.

Research Foundations

The special nature of foundations in the United States is characteristic of American fiscal policy. The international impact of these entities in the agronomic sector was shown by the International Rice Research Institute in the Philippines and the international grain institute, CIMMYT, in Mexico created by the Ford and Rockefeller foundations. These two institutes gained renown for their discoveries of high-yield dwarf varieties of rice and wheat, which won the Nobel Peace Prize for the agronomist Norman Borlaug.

Because of these results, the World Bank, at the direction of its former president (1968–82), Robert MacNamara, began to devote greater attention to agricultural and rural problems. More than one-third of the World Bank's recent budgets, which have increased fivefold since the arrival of MacNamara, is now allocated to agricultural and rural projects.

In association with the FAO and the UNDP (United Nations Development

Program), the World Bank wanted to extend the action of the foundations by organizing a network of international agricultural research centers now composed of twelve specialized centers financed by around thirty countries and organizations which voluntarily contribute more than $100 million every year. The overall budget of these twelve centers increased from $20 million in 1972 to $118 million in 1978 and to $140 million in 1980.

A study by the World Bank showed that the annual growth rate for yields is well correlated to the percentage of the arable land area cultivated in dwarf varieties.[20] Thus, India and Pakistan, which planted approximately 60 percent of their land with high-yield varieties in 1973, experienced an annual growth rate of 5 percent of their wheat yields from 1965 to 1973. Indonesia, which planted 40 percent of its land with dwarf varieties of rice in 1974, enjoyed an annual yield growth rate of 3 percent from 1965 to 1973. Since 1970, many multinational companies have signed exclusivity agreements for the production of seeds certified as high-yield varieties.[21]

These twelve international centers direct most of their efforts toward vegetables and livestock and have not entered into the area of so-called "industrial" tropical crops, in which the nine French GERDAT-CIRAD centers (international centers for agricultural research and development) enjoy a substantial technological lead in rubber, cotton, oilseeds, coffee, and tropical and citrus fruit.

Foundations and associations also play an eminent role in economic research. Organizations such as the Brookings Institute, the Atlantic Council, the Overseas Development Council, the Conference Board, and the American Enterprise Institute for Public Policy Research regularly conduct collective studies or retain agricultural policy specialists, often former cabinet members (political appointees), for the preparation of updates or background studies.

As often is the case in the United States, it is not the amount of funds, particularly public funds, which is behind these successes, but rather the originality and compatibility of the structures. They allow much faster circulation and dissemination of research results as well as a better research orientation and reorientation. These structures spread the notion of national interest far beyond the political sphere, to which it is normally limited in Europe. They have allowed the periodic revision of agricultural policies by preparing the national consensus necessary for any reform in substance or style in this fundamental American sector. They therefore complement the public financial resources, which will be studied next from both internal and external viewpoints.

LOANS AND LAND OWNERSHIP

Land tenure and land ownership policy do not appear to be among the instruments of American agricultural policy, at least not per se. In an economy in which land represents 75 percent of total agricultural assets, however, one must naturally look to credit policies to find the true instrument of land redistribution. The increasing difficulties beginning in 1983 of some middle-size farmers to pay back their loans and the consequences for their banks and for the Farm Credit System are giving the farm debt problem a new importance.

Types of Loans and Lenders

There are three principal forms of agricultural credit: (1) *production loans,* which can be either short or medium term and are used for the purchase of feed grain, livestock, milk cows, or machinery, or to finance storage; (2) *real estate loans,* which are long term and intended for the purchase or extension of farmland, financing for farm buildings, or refinancing; and (3) *loans to cooperatives,* which are used to cover their operating costs and to finance their stocks, construction, or building purchases. These three types of loans are representative of the specificity of agricultural credit demand. Farming is an activity that depends highly on weather conditions and the rhythms of the seasons, requires numerous and expensive production inputs (such as fertilizer and machinery), and involves a constant process of land property transfers.

Lenders are classified, by almost unanimous agreement, in two groups: those that make short- and medium-term loans for operating capital and those that make long-term real estate loans.[22]

Operating Loans

Commercial banks are the principal sources of operating loans, although their share has been steadily decreasing since 1975. Historically, they have played an important role because of their eagerness to provide small short- or medium-term loans. Another factor reinforcing the role of commercial banks is the great diversity among farmers. As emphasized by J. Hand, this explains the important role of personal relations between farmer and banker. "The professional capabilities of the farmer as a technician, businessman and manager, are of crucial importance, but can be appreciated by a potential lender only through personal, direct contact."[23]

The Farm Credit System is a mutual credit system created in 1916 by the Federal Farm Loan Act "for the purpose of centralizing all federal government activities relating to agricultural credit in a single agency." The twelve Farm Intermediate Credit Banks, which officially constitute "a permanent, reliable and independent source of short and medium term agricultural credit," are controlled and directed by the Farm Credit Administration.[24] They operate via the 425 Production Credit Associations, which are in direct contact with the farmers, who are both their clients and their owners. The interest rates charged by these two agencies are identical, regardless of the solvency of the borrower, but are generally higher than those charged by commercial banks. The commercial banks can adopt a more flexible policy, varying their interest rates based on the collateral furnished by the client. As a result, large farm operators approach the commercial banks first.

A third category is "individuals and others." This category encompasses two types of loans. First, wholesale merchants, businessmen, and the like provide certain farmers with short- or medium-term loans. They then discount

TABLE 6.8. Contribution by the various types of lenders to agricultural production credit

	Total Outstanding Credits in Billions of Dollars	Commercial Banks (%)	Farm Credit System (%)	Farmers' Home Administration (%)	Private Individuals and Others (%)	CCC (%)
1965	17.91	39.0	13.4	3.6	35.4	8.6
1970	23.84	43.3	19.8	3.3	22.0	11.2
1975	35.55	51.3	27.9	2.9	17.0	0.9
1980	75.77	41.0	25.2	11.4	15.5	6.7
1982	92.81	35.5	24.1	15.6	16.2	8.6
1983	106.60	33.9	19.6	13.6	18.5	14.4
1984	103.10	37.3	19.1	15.1	18.3	10.1

Source: USDA, Agricultural Statistics, 1984, Table 675, "Non Real Estate Farm Loans, Amount Outstanding by Specified Lender."

the bills of exchange with commercial banks or other financial institutions. Next, there are private individuals whose role as lenders is considerably less important than their roles in financing real estate loans, as will be shown further on.

Last but not least comes the Farmers' Home Administration (FHA), created in 1946. This agency, a division of the USDA, conducts a program of insured loans for smaller and younger farmers unable to obtain credit from other sources. According to Ludke, "these loans are primarily used for the development and expansion of family farms," which have not yet reached a commercial level.[25] Interventions by the FHA have recently taken on special significance because of the growth in advantageous loans for disasters (the Economic Emergency Loan and Emergency Disaster Program of 1983). Outstanding loans by the FHA thus increased from $0.8 billion in 1971 to $8.9 billion in 1979 and to $15.6 billion in 1984. This reflects the crisis of middle-size farms as publicized by Country, a 1984 movie showing an Iowa farm bankruptcy.

Finally, the Commodity Credit Corporation participates in loan activities through its price support operations, crop year loans, and three-year farmer-owned reserve loans. The evolution in credit sources is summarized in Table 6.8.

Real Estate Loans

Real estate loans are made by a number of institutions already mentioned. Traditionally, individuals, usually fairly close relatives of the farmer, represent the largest source of credit. Far behind are commercial banks and the Farmers' Home Administration. On an equal footing with individuals, once again is the Farm Credit System, represented by the twelve Federal Land Banks. They are

TABLE 6.9. Contribution of the various creditors to real estate loans (in percent of total outstanding land credits)

	Total Outstanding Credits in Billions of Dollars	Federal Land Banks and Associations (%)	Farmers' Home Administration (%)	Life Insurance Companies (%)	Commercial Banks (%)	Private Individuals and Others (%)
1965	18.89	19.5	6.8	22.7	12.8	38.2
1970	29.18	22.9	7.8	19.6	12.1	37.5
1975	46.29	29.0	6.9	13.6	12.9	37.6
1980	82.68	35.8	8.6	15.2	10.0	30.4
1984	111.90	43.0	8.5	11.4	8.1	28.9

Source: USDA, Agricultural Statistics, 1984, Table 671, "Farm Real Estate Debt: Amount Outstanding by Lender."

responsible for procuring a permanent source of mortgage loans for farmers at reasonable rates through 505 Federal Land Bank Associations. Surprisingly, insurance companies are the third largest source. Their activities in this area have developed in recent years, primarily in the form of loans to very large farms. Table 6.9, which is similar to that prepared for operating loans, allows the role of the various creditors to be viewed from a historical perspective.

Another interesting aspect of the mortgage loan market is the average amount of the loans provided by each category of lenders. Insurance companies provide loans more than twice as large as those furnished by specialized agricultural banks.

Changes in Farm Financial Balance Sheets

The proportion of debt to assets has increased considerably since 1950. That year it was 10 percent; it increased to 13 percent in 1960 and to 20 percent in 1970, stabilizing at that level during the 1970s.

Liabilities for the agricultural sector increased from $12.4 billion in 1950 to $53 billion in 1970, to $140 billion in 1979, and exceeded $215 billion in 1984. The average debt per farm increased from $2,200 in 1950 to $18,000 in 1970, to $38,000 in 1977, and to $79,000 in 1982. Of this total, slightly more than 50 percent are real estate loans. American agriculture, therefore, depends increasingly on credit, an evolution that is probably the result of the structural transformation of the sector. Further examination of the balance sheet leads to other conclusions.

At first glance, it would appear that farmers are willing to go into debt to purchase land. Indeed, debt in the form of farm credits increased by a factor of 4.1 between 1950 and 1970, while real estate debt increased by a factor of 5.2. This statement, however, should be slightly qualified. Traditionally, farmers like to reinvest profits in their own operations. This was particularly true during the reference period: financial assets in the form of outside investments amounted to 12 percent of total assets in 1950 and fell to 5.7 percent by 1970. During the same period land prices rose much more rapidly than those for machinery and vehicles.[26] What can we conclude from these figures? First, farmers do not like to resort to outside financing, but they are doing so increasingly. Second, they have reinvested their profits to a massive degree in their own operations, and they have used credit essentially for the purchase of equipment.

The 1970s present the opposite picture. Early in that decade, from 1972 to 1974, the sector's assets increased by $150 billion, while liabilities increased by only $20 billion. During the second half of the decade, however, this trend was reversed. From January 1, 1976, to January 1, 1978, assets increased by 22 percent and liabilities increased by 33 percent.

This situation is not alarming because the ratio between assets and liabilities remained very reasonable compared with other sectors of the national economy except for the 7 percent of farmers whose liabilities exceeded 70 percent of their assets in 1984 (2 percent in the 1970s). The problem raised is connected with the additional obligation incumbent upon farmers who must reimburse a higher principal. Combined with decreasing income and higher real estate taxes, this has resulted in a drop in yield on capital invested in agriculture. Yield, which had increased from 5.1 percent in 1971 to 10 percent in 1973, declined to 2.3 percent in 1977. Another way to measure this phenomenon is to consider that if all net agricultural income had been channeled toward eliminating liabilities, all of the liabilities of the agricultural sector would have been eliminated in 2.2 years in 1973, in 5.1 years in 1977, and in 7 years in 1984.

These figures, however, hide one of the most fundamental aspects of the evolution in American agriculture: the extremely unequal distribution of these liabilities among farmers.

Debt as an Input

Credit has played a considerable role in restructuring the agricultural sector. This can be verified simply by observing that the debt/asset ratio varies considerably according to the category of farmer (see Table 6.10). In 1982 it varied from 32 to 13 percent because most of the farmers in categories 1, 2, and 3 were in debt, compared with less than half of those in category 7. When these figures are compared, it can be seen that farmers in categories 1 to 4 constituted 29 percent of the entire population but accounted for 78 percent of total debt. The Lorenz distribution is even more unequal when average debts per farmer are compared: $1,373,000 for farmers in category 1, or 2.4 times their annual income, compared with $173,000 for those in categories 2, 3, and 4, or 7 times their annual income. This explains why the 180,000 farmers whose debts exceed 70 percent of assets are almost all in categories 2, 3, and 4 (sales from $40,000 to $500,000). One out of four farmers in these categories is potentially bankrupt. The farm banks represented 36 percent of all problem banks in

TABLE 6.10. **Ratio of debt to assets, to total income, and to gross receipts, 1982**

Category in Sales According to the Farm's Class	Number of Farms (%)	Assets/ Farm	Debt/ Farm	Debt/ Assets %	Total Income per Farm	Debt/ Income	Gross Receipts/ Farm	Debt/ Receipts %
1. More than $500,000	1.0	4,247	1,373	32	575	2.4	1,790	77
2. $200,000–$499,999	3.6	1,760	398	23	63	6.3	325	122
3. $100,000–$199,000	7.8	1,077	206	19	29	7.1	154	133
4. $40,000–$99,999	16.3	624	108	17	15	7.0	73	148
5. $20,000–$39,999	11.4	360	52	14	13	4.0	33	158
6. $10,000–19,999	11.8	218	31	14	16	1.9	17	186
7. $5,000–$9,999	13.8	142	19	13	18	1.0	9	217
8. Less than $5,000	34.3	85	12	14	20	0.6	3	410
Average for all farms	100.0	408	79	19	26	3.1	62	126

Source: Economic Indicators of the Farm Sector (Washington, D.C.: ERS, USDA, 1983).
Notes: Units: thousands of dollars or % of one unit. Total number of farms in 1982: 2.4 million. The total income constitutes the net farm income and the off-farm income. The gross receipts are exclusively agricultural.

1984 (only 22 percent in 1983). The whole system of farm banks is in jeopardy, and the 25 percent drop in farmland prices that occurred between 1981 and 1984 has decreased the securities of the banks, which thought they were guaranteed by the sales of farmland.

According to a survey in 1970, the yield on capital invested was 6.9 percent for categories 1 to 3; 5.9 percent for category 4; 4.4 percent for category 5; and 2.9 percent for category 6. From category 7 on, yields were negative ($-.1$ percent for category 7 and -6.3 percent for category 8). In the 1980s yields have become negative for all categories except 1 to 4 (see Appendix 9).

There are more reasons behind the impressive performance by large farms and the drama of the middle-sized farms. First, large farms use less capital to generate the same income ($2.64 in assets per dollar of gross income in category 1 compared with $5.10 in category 2, $8.50 in category 4, and $12.80 in category 6). The second explanation relates to the fact that for large producers the ratio of total assets to equity is much higher, since they borrow to finance operating capital rather than to finance real estate. In 1982, farmers in categories 1 and 2 represented 5 percent of the population, yet they accounted for 50 percent of the sector's total sales and more than one-third of its total debt.

In conclusion, it appears very clear that large farms can resort much more easily to credit than can middle-sized farms and that investments made with this credit yield more than they would have on middle-sized farms. The result is a land policy that encourages the formation of large farms with sales greater than $20,000 in 1970 or $100,000 in 1982.

GOVERNMENT ASSISTANCE FOR AGRICULTURAL EXPORTS

The market of these large but still family-managed farms is more and more export-oriented. A government policy in this field becomes essential. As seen in Chapter 4, food aid, progressively substituted by a policy of medium- and long-term credit, played a great role in the absorption of American food surpluses. Since 1966, however, different means, tending to be less interventionist, have been used to support American agricultural markets and exports. PL 480 has been reoriented, the celebrated sale of American products for local currencies has been replaced by long-term credit, payable in twenty or even forty years, and pure grants declined until 1975. The following section will examine the reasons for this change and the trend toward greater export aid that has emerged since the great debt crisis of the developing countries beginning in 1981.

PL 480

The Agricultural Trade and Development Assistance Act of 1954, passed under the Eisenhower administration and better known as Public Law 480, initially responded to several needs expressed in the following order:

1. Facilitate currency convertibility.
2. Promote the stability and prosperity of American agriculture.

3. Use American agricultural surpluses to stimulate the expansion of international agricultural trade.

4. Encourage the economic development of foreign countries.

5. Procure the financial means to purchase strategic materials and to pay for American overseas expenditures.

6. In short, reinforce American foreign policy.

In 1966, however, the Food for Peace Act (Public Law 89-808) profoundly changed the goals of PL 480, the name still used for this program. The Democratic administration, still under the influence of the Kennedy team, wanted to apply the results of twelve years of experience in which stocks of wheat estimated at 1.5 times the domestic consumption of the United States in 1954 increased in 1961–63 to 2.3 times domestic consumption, then declined to 1.1 in 1965 and 0.8 in 1966 after reduction in the support price level. References to American surpluses, therefore, became unnecessary. During the same period, Europe enjoyed economic recovery and no longer needed food assistance. The American balance of payments began to pose a problem. The nonindustrialized countries, meanwhile, particularly the Latin American countries, had begun advocating the need for world growth in favor of development.

The objectives for 1966 were thus outlined in the following order:

1. Stimulate growth in world trade.

2. Develop foreign markets likely to import American agricultural products.

3. Use the "abundant productivity of American agriculture" to "combat hunger and malnutrition" and to "encourage the economic development" of "developing" countries. Food assistance must be granted on a priority basis to countries engaged in an effort to develop local agriculture.

4. Promote by other means the foreign policy of the United States.

In 1972, finally, discussions took place within the Nixon administration to reorient and rationalize American civil and military foreign aid. The result of these discussions was a number of basic principles concerning the application of PL 480:

1. The law should be used to increase commercial sales of agricultural products, encouraging programs to develop local animal husbandry, for example, so as to create demand for feed grain and soybeans.

2. Limit pure grants to charitable and emergency activities, and, if possible, channel this aid through international institutions such as the World Food Program or the United Nations Refugees of War Agency.

3. Replace sales in local currency with long-term loans partially payable in local currency with interest rates based on the borrowing country's capacity to repay (the 1966 law prohibited sales in local currency effective in 1972).

In 1975, following the oil crisis and the massive sales to the Soviet Union, and after Watergate had allowed Democratic congressmen, particularly Senator Humphrey, to play a central role in agricultural policy, the new food aid law emphasized the need to assist the poorest countries on a priority basis: 75 percent of sales involving long-term credit had to be reserved for countries with annual per capita incomes below the poverty threshold, that is, less than $300 (increased to $580 in 1979 and $805 in 1985).

Furthermore, the commitments made during the World Food Conference, held in Rome in 1974 under the initiative of Henry Kissinger, caused the United States to increase considerably the level of aid, distributed as follows:

1.3 million metric tons in grain equivalents in 1976
1.6 million metric tons in grain equivalents in 1978
1.7 million metric tons in grain equivalents in 1982
3.0 million metric tons in grain equivalents in 1985

The 1977 farm bill introduced a new concept: the Food for Development Program. Beginning in 1979, 15 percent of long-term credit sales would be assigned to countries that could channel annual loan payments owing to the American Treasury toward their own rural and agricultural development projects, which especially favored small farmers (selected seeds, fertilizers, agricultural equipment, roads, irrigation, electrification, storage, research, and extension). For example, a project in Bolivia involved the construction of cooperative silos, agricultural supply centers, and irrigation, financed with $75 million over five years.

On the thirtieth anniversary of PL 480, July 10, 1984, President Reagan announced a series of measures inspired by the Task Force on Hunger, headed by Robert Keating, U.S. ambassador to Madagascar. These proposals are designed both to improve food exports during a time of increasing surpluses (30 percent of food and agricultural exports from the United States are sent to developing countries), to relieve the financial difficulties of the increasingly indebted developing countries, and to increase these countries' own production capacities. President Reagan seems to have been influenced by the meetings in Cancun with the leaders of the Third World and in Alaska with Pope John Paul II in May 1984, as well as by Republican Senators Robert Dole and John Danforth.

The announced measures completed on January 3, 1985, under the name Food for Progress, deal with the creation of security stocks in Africa, an emergency food aid fund of 1.5 million tons of grain, the possibility of financing the land and sea shipping costs for the aid, and the reinforcement of the food shortage forecasting system. There are also plans to study longer-term measures directing assistance toward the development of local production in the developing countries and the creation of intermediate programs—that is, between aid and cash sales.

The principal innovation contained in PL 480, in contrast with the traditional food assistance systems, is not solely the diversity of its objectives. Its methods of finance and the increasing use of long-term soft loans are also innovative.

The grants amounted to more than $6 billion over twenty years, from 1955 to 1975, excluding the packing and transportation expenses, which amounted to nearly $2 billion. The main products provided as aid during famines and natural catastrophes, or to combat permanent infantile malnutrition, were, in order of declining quantity: powdered milk, flour, wheat, soy oil, and, most recently, new protein-enriched products such as CSM (corn-soy milk) or WSB (wheat-soy blend). These donations provided material support to certain U.S. AID development programs and procured more than 40 percent of the resources of the United Nations World Food Program.

Grants represent only 22 percent of exports under PL 480, however, because long-term credit sales played the most important role after 1966 (70 percent of the value of products exported over twenty years).

These special term sales fall into two categories:

1. Sales with credit terms of forty years (a ten-year grace period with interest at 2 percent, followed by thirty years at 3 percent) for 95 percent of the value of the product sold. Repayment is made partially in the local currency if American embassies are in a position to use it locally. With "normal" interest rates of 10 percent per year, this type of credit corresponds in reality to a subsidy of 70 percent of the value of the merchandise as expressed in world prices.

2. Sales with credit terms of twenty years (a two-year grace period with interest at 2 percent, followed by eighteen years at 3 percent) for 95 percent of the overall value. Repayment is made exclusively in dollars. Given the same rate of inflation, this type of loan corresponds to a subsidy of about 44 percent.

The law requires that 50 percent of this merchandise be transported on American ships, which obliges the Commodity Credit Corporation, the financial agency within the American Department of Agriculture that administers PL 480, to subsidize American companies, whose costs are higher than those of foreign competitors. This assistance to shipping companies represents about 8 percent of the value of the products.

Credit sales over twenty years (1955–75) have amounted to more than $18 billion involving the following products: wheat and flour ($9.7 billion), rice ($2.6), cotton ($2.4), soy oil ($1.6), corn and sorghum ($1.2), tobacco ($0.6), and dairy products ($0.2). In addition more than $6 billion was spent for packing and transportation assistance.

The barter system has not been used since 1970 and represented only 8 percent of total food assistance credits. Europe and Japan have been the principal beneficiaries.

Food assistance, particularly from the United States, the world's largest provider in both value and variety, poses a number of political problems for the beneficiary countries and a number of economic problems for nations competing with the United States. A few little-known figures are worthy of greater dissemination.[27] It could thus be observed that the home of free trade, the United States, in 1976 still exported 26.5 percent of its rice thanks to the aid program and 22 percent of its cotton and soy oils. Although the importance of PL 480 for wheat exports has decreased considerably, from 65 percent in

1961–65 to 45 percent in 1966–70 and to 10 percent in 1971–75, in 1976, 8.5 percent of American wheat production was still exported through PL 480. To be even more precise, in 1970, 87 percent of wheat flour, 67 percent of condensed milk, 97 percent of powdered milk, 69 percent of cornmeal, 97 percent of bulgur, and 89 percent of CSM–WSB (a mixture of corn and soy or wheat and soy in the form of flour) found an outlet through PL 480. The major portion of cotton, soybean, corn, and sorghum, however, has been sold in the 1980s through mechanisms that do not resort to food assistance procedures.

This evolution has been made possible because commercial exports from the United States to a large number of countries have been developed. For example, Japan, Spain, Brazil, Taiwan, Iran, Peru, and Colombia, after benefiting from PL 480, all became faithful clients for American agricultural products. The same is true of India, Korea, and Egypt, which all increased their commercial imports from the United States. The commercial objectives of PL 480, therefore, seem to have been gradually attained, which enables the United States now to channel its aid toward new markets in Africa and Asia. Thus the United States has a wide range of aid possibilities, from pure grants to sales at 56 percent of the world rate (twenty-year loans) to 30 percent of world rates (forty-year loans).

The law is administered so as to create new, solvent, and permanent foreign markets based on the needs and financial possibilities of the developing countries. The two following examples illustrate this intention. The sale on credit of wheat and soy oil to a private bank in a Middle Eastern country made it possible, with the income from local resale, to finance the completion of a production unit for 7,300 metric tons of chickens per year, which will enable the country to become a purchaser of soy, grain, and American day-old chickens. The sale on credit of 14,000 metric tons of corn to a company from a Caribbean island will allow the construction of a corn–soy meal plant. Through these expenditures, Italy, Spain, and Japan in the 1950s and 1960s became consumers of American grain. Through PL 480, hamburgers became a "national" dish in Japan, with McDonald's becoming the principal seller.

The forty-year and later the twenty-year credits allow a transition toward cash sales (or at least with a three-year credit normally provided by the Commodity Credit Corporation). Interest rates are progressively increased, and the repayment periods are progressively reduced.

The advantages of sales based on credit for the U.S. balance of payments are considerable. It has been calculated that in 1970, when spending on soft loans rose to $743 million, the U.S. balance of payments recovered $328 million: (1) $225 million that had been spent by the embassies and American services overseas was recovered in local currency by repayment of the forty-year loans; (2) $103 million was repaid in dollars for previous contracts.

PL 480 was created and has been used as a particularly effective tool of U.S. foreign policy. Thanks to the receipts in local currency which it enables the United States to recover, it has facilitated, despite modifications, the maintenance of powerful embassies supported by propaganda services such as the

U.S. Information Service, AID, or direct action agencies such as the CIA. During the Korean and Indochina wars PL 480 reduced the official cost of maintaining the expeditionary force.

Indeed, because of PL 480, the financing of foreign activities is indirect and takes place with little control by Congress. During the postwar period, sales in local currency allowed the United States to finance at small cost the imperial expenses the United States was incurring in Western Europe and Japan.

Many economic expansion programs were thus made possible. The following examples are from the year 1971. The promotion of soy, deboned chicken, and practically all American products was eligible for partial financing: 50 percent by the government, in part with local PL 480 funds ($138 million were used from 1954 to 1971) and 50 percent by American professional associations. Market studies costing $750,000 in 1971 and food technology studies for American products costing $500,000 the same year all had the same nonbudgetary source. The secretaries of health ($32 million in 1971), agriculture ($54 million), commerce ($384,000 for the environment), state ($10.7 million for American schools overseas), the Smithsonian Institution ($2.5 million), the Library of Congress (to purchase foreign reviews), and the National Science Foundation ($5.3 million) all benefited from the Food for Peace program. The list of American accomplishments in the developing countries resulting from food assistance income is still longer. Construction costs for civil buildings amounted to $41 million over twenty years and those for military buildings reached $90 million. Purchases of military equipment by countries receiving aid under the Food for Peace program in 1971 were evaluated at $78 million for the purchase of uniforms in Vietnam, $20 million for coastal defense in Korea, and $655,000 for Cambodia. American businesses with overseas operations benefited in 1971 in the amount of $5 million for a livestock feed plant, a nylon factory, and a hotel. Economic development programs were financed: a loan to India of $64.4 million and subsidies for Pakistan of $51.6 million. Family planning operations in India and Ghana and the financing of Taiwanese technical assistance in Africa, Latin America, and Asia also came from PL 480. The remaining repayments in local currency were used by American tourists ($2.6 million).

American food aid is indeed a powerful instrument of foreign policy, used, notably, in Southeast Asia (Korea, Indochina). More recently aid was granted to Egypt, Israel, Jordan, Syria, Tunisia, and Morocco in a move obviously linked to President Carter's Middle East peace plan. Appendix 17 shows that in 1979–80 the Middle East–Maghreb received 35 percent of the total assistance granted by the United States, without counting Sudan and Somalia, which each received 4 percent of the credits granted. The rest of Africa received only 7 percent of the total and Latin America, 11 percent. Portugal, incidentally, is the only European country that received food assistance in 1979–80.

Food aid is also a commercial tool that links the recipient country to American supplies. This is one reason why India decided in 1976 no longer to resort to credit sales with special conditions from the United States. In 1965, 20 million Indians had been fed for a year by such sales. Not only does the recipient country become bound for twenty or forty years to the donor country, which it must repay annually the amount stipulated in the contract, but also it must respect a certain number of constraints for the life of the contract concerning foreign sales of the products accepted as a donation. The technical name of these constraints should not dissimulate the substantial commitment they represent. They are as follows:

1. The Usual Marketing Requirement (UMR), which involves a commitment by the recipient country to continue purchasing identical quantities on the commercial market as those purchased prior to the supply of American aid.

2. The Tied-UMR, which involves the purchase in the United States of a quantity identical to that preceding the year of aid.

3. The Tied Offset Requirement: If the recipient country transforms the product received or an analogous product and/or reexports the transformed product or a product similar to the product received, it must agree to repurchase, on a commercial basis, amounts equal to the amounts exported. This repurchase must take place in the United States.

An intelligent food aid policy can thus help in the development of international trade. When such a policy features credits with special conditions, however, it can plunge the recipient country into a situation that binds it financially and commercially to the donor country.

Another criticism of food assistance relates to the detrimental effect of American aid on the production of local farmers. The prices for their products are forced downward and governments have a tendency to lose interest in their plight, since their large cities receive abundant provisions at low prices.[28]

The American government and operatives quickly became aware of the difficulties these political and commercial constraints were causing for the beneficiary countries. They realized that this food aid policy could provoke opposition throughout the world. That is why, as early as 1956, they attempted to formulate a new credit policy to facilitate the purchase of agricultural products by countries that would have difficulty paying cash. This credit, granted almost automatically based on bank guarantees alone, was to allow the development of American trade while avoiding the criticisms building against PL 480.

Medium-Term Export Credits from the Commodity Credit Corporation

Created during the New Deal in 1933 by President Roosevelt, the CCC is the financial arm of the U.S. Department of Agriculture for the organization of agricultural product markets. In reality, this company is simply an administrative board chaired by the undersecretary of agriculture and is entirely managed by an office within the department called the Agricultural Stabilization and Conservation Service.

Beyond its price support activities conducted through the financing of stocks held by American farmers, the CCC administers two other programs directly linked to export operations: PL 480 and medium-term credit. Out of a

1978 budget with total expenditures of about $6 billion, $1.2 billion was devoted to PL 480 and $1.6 billion to medium-term credits, the remainder being used to support domestic markets. Repayments of previous loans, however, helped to compensate for expenditures of $384 million under PL 480 and of $640 million under medium-term export credit (see Table 6.11).

The use of these short- and medium-term export credits (up to three years) created in 1956 has expanded to replace both export subsidies and food assistance. In 1965, out of $6 billion in exported agricultural products, 26 percent was exported because of PL 480, 29 percent because of export subsidies, and only 1.5 percent because of three-year credits. In 1973, the year of the great sales to the Soviet Union, out of $11 billion in exports to the whole world, only 9 percent was because of PL 480, 28 percent because of export subsidies, and 19 percent because of two- and three-year credits and financing procedures implemented to favor compensation operations in the Eastern bloc countries. In 1978, out of $29 billion in exports, PL 480 accounted for only 4 percent and medium-term credit for no more than 6 percent.[29] The development of overseas sales, which followed the Soviet purchases in 1973, occurred normally without any buyer's credit or export subsidies. (See Appendix 18.)

The evolution that has occurred is therefore considerable, since export subsidies had been granted until 1973, the last year in which they were used on a large scale, for products representing 27 to 35 percent of total exports depending on the year. The turning point came in 1974 under the Republican administration. Agriculture Secretary Earl Butz and his assistant Carroll Brunthaver, former (and future) director of market studies for Cook, decided, after the scandalous 1972 Soviet grain deal, to eliminate export subsidies and rely exclusively on medium-term credit (one to three years) for American agricultural exports. Congress, however, was cautious enough to leave export subsidies in the law, and they still remain. Yet despite pressure from the agricultural lobby and the prospect of further accumulation of stocks, Robert Bergland, agriculture secretary in the Carter administration, did not reverse the moral commitment made by his predecessor, one of the most popular secretaries since World War II. It was not until the Reagan administration, under Secretary of Agriculture John Block, that direct export subsidies were reestablished, with subsidized sales to Egypt and blended credit, one-fourth of which consisted of no-interest loans called GSM 5.

Established as the prime instrument of export policy, medium-term credit amounted to $1.5 billion in 1979. The overall amount authorized for outstanding loans by the CCC was increased to $24 billion in 1984 by a U.S. Treasury decision (in mid-1978 this amount was $10.7 billion, of which $1.8 billion was intended for medium-term export credit).

The mechanism is simple: any American exporter or any importing country can submit a request to the CCC for a loan for six months to three years by submitting an irrevocable letter of credit, guaranteed by an American bank. A loan of 100 percent of the FOB value exported is then automatically granted to the foreign importer. Its rate is calculated each month by the American Treas-

TABLE 6.11. CCC expenditures and receipts in fiscal year 1978–1979 (millions of dollars)

	Support of Domestic Markets	PL 480	Medium-term Export Credits	Total per Product
Feed grains	765	99	525	1,389
Wheat	1,195	595	468	2,257
Rice	3	162	17	182
Cotton	85	10	230	326
Vegetable oils	5	138	39	181
Oilseeds	71	21	244	336
Dairy products	193	27	0	220
Sugar	212	0	0	212
Total spending	3,056	1,192	1,583	5,864
Total receipts (repayment of prior loans)	0	−384	−640	−1,024
Net spending	3,056	808	943	4,840

Source: USDA, ASCS, *Report of Financial Condition and Operations, CCC*, September 30, 1978.

ury based on the average cost of its funds over one year. This rate was 7.375 percent on May 1, 1978, but it climbed to 20.5 percent by April 7, 1980.

If the letter of credit is guaranteed by a foreign bank, 10 percent of its amount must be guaranteed by an American banking institution. In this case, the interest rate is increased by .125 percent. This system is similar to that of a "buyer's credit" granted automatically to all American exporters of agricultural products appearing on a list published by the U.S. secretary of agriculture. It allows agricultural exports to benefit from interest rates used for American Treasury bonds.

The accumulation of grain stocks in the United States has induced Congress to place renewed emphasis on this medium-term credit program. The Agricultural Export Trade Expansion Act of 1978 allows the CCC to offer medium-term credit (GSM 201 and 301), which may exceed three years in duration for the following products and facilities: (a) products covered in international agreements involving an international system of reserves; (b) breeding livestock, including transport costs (five years at most for cattle, hogs, sheep, poultry, and the like); (c) facilities in the importing countries designed to facilitate the transport, unloading, stockpiling, processing, and commercialization of agricultural products or livestock; (d) any agricultural products for which competing countries are offering longer-term financing, endangering American exports.

Despite the interest in intermediate-term loans of three to ten years (Intermediate Credit Program GSM 201 and 301), such loans have not exceeded $300 million per year in 1978–85. Preference has been given to extension beyond the three-year-loan term on GSM 102 when the country has financial problems in repaying its loan.

The same legislation now enables American exporters to receive loans for three years in the form of "supplier's credit," whereas the previous system limited these loans to foreign importers with irrevocable guaranteed letters of credit. Finally, the law extends the benefits of the six-month to three-year loans to the People's Republic of China, but, under the Trade Act of 1974, only Poland, Romania, Yugoslavia, and Hungary, among the Eastern bloc countries, may receive these benefits. The Soviet Union has therefore been excluded since 1973 from this financing.

Under the Reagan administration, medium-term credits of six months to three years were replaced by a CCC guarantee system applied to loans from private banks. This system, called GSM 102, applies to $5 billion for fiscal year 1985. It makes the government more a guarantor than a direct lender. To pose a threat to the EEC, blended credits of $2.5 and $6 billion respectively in 1983 and 1984 were developed, combining 75 percent GSM 102 (three-year loans at commercial rates) with 25 percent GSM 5 (three-year loans with no interest). This formidable but costly weapon was used on Morocco, Brazil, and Egypt. This was not considered sufficient to promote exports and on May 15, 1985, a new program of export subsidy called BICEP (Bonus Incentive Commodity Export Program) or GSM 500 was designed and granted $500 million. Exports of millions of tons of wheat or flour have been offered to Algeria, Egypt, and Yemen, with an export bonus of $21 a ton of wheat. The trade war in wheat with EEC is open.

The financial mechanism made available to American exporters is thus nearly complete, ranging from six months at the commercial interest rate to forty years at 2 percent, but a return to export subsidies has been started in 1982, which is a change from policy started in 1960 to unify world prices and American prices. During this period of rapid growth in the debt of developing countries, which began in 1981, it is natural to see the share of American food and agricultural exports supported by government credit increase, which it has—from 9 percent in 1981–82 to 18 percent in 1982–83. This accentuated trend toward greater aid to exports is one of the main points of the 1985 farm bill to compensate for the decrease of support prices and the freeze of target prices. (See Appendix 20.)

Export Promotion

This credit assistance mechanism, however, was not enough without a special effort to promote American food, feed, and fiber abroad. The free enterprise philosophy that prevails in the United States has long dissuaded the American government from becoming directly involved in the sale of agricultural and food products overseas. Its traditional role has therefore been the circulation of economic information on foreign production and markets, using the remarkable network of American agricultural attachés.

Constituted in May 1919, agricultural attachés numbered 110 in 1979, assisted by 90 foreign professionals. They are located in seventy countries with staffs ranging in size from one professional and his secretary to offices headed by an agricultural counselor (the 1978 law created ten such positions for the principal U.S. clients and competitors). The offices often include two or three agricultural attachés, of American nationality, and two or three assistants who are nationals of the host country.[30]

What is most striking about the activities of these offices is the organization of their information-gathering systems. Two offices within the Foreign Agricultural Service (FAS) of the USDA, the Office of Agricultural Attachés and the Office of Economic Analysis, continually organize the dissemination of economic information assembled at their request by the embassies. The information is communicated through the use of standardized questionnaire file cards with well-defined dispatch dates. Because of this system, projections of world production of the various agricultural commodities published by the FAS foreign circulars are generally available more rapidly than those made by the various international organizations (World Wheat Council, FAO, and the like) or by the many independent or bank economic services.

It is the most extensive system of its kind in the world. The Netherlands, the world's third largest exporter of food and agricultural products, maintains a system that covers only twenty-eight countries. The system in France, the world's second largest exporter, covered only eight countries in 1974 and only twenty-five in 1980. Yet the cost of the American system is relatively low: $10 million for the attachés and their Washington correspondents and $3 million for the Office of Economic Analysis (1979 basis).

Another result of the effort of the agricultural attachés in the area of information is the Trade Opportunity Referral Service (TORS). This service, created in 1972, is designed especially to inform the small-scale food industries of foreign sales possibilities. Using a computer, this service matches foreign purchasers of agricultural products and American suppliers. More than three thousand American businesses are enrolled as correspondents on the list of foreign purchase offers transmitted to the central office by agricultural attachés stationed overseas. The annual cost for TORS is relatively modest. The development of a computer program amounted to about $200,000, and operating costs are only about $50,000 per year for a resulting sales figure estimated in 1978 at $15 million.

The American approach has evolved in recent years, and the USDA has progressively developed, in collaboration with professional agricultural and food organizations, a system of promoting American products overseas.

Actually, as early as 1954, using credit in local currency obtained through PL 480, the embassies began to finance propaganda campaigns for American food. The most famous of these programs was the introduction of the hamburger to the Japanese market. Intended to develop ground beef consumption in Japan, the program resulted in a massive infusion of American wheat into Japan for the production of hamburger buns. Once exclusive rice consumers, the Japanese became rice exporters and wheat importers.

The system put into place by the United States for these promotions is original because it has not given rise, as it has in Europe, to the creation of quasi-governmental corporations such as the CMA in Germany or the SOPEXA in France. Multiyear contracts have been signed by the Foreign Agricultural Service with forty professional agricultural organizations. These organizations, called Market Development Cooperators,[31] are themselves financed jointly by nine thousand American exporting companies, fifteen hundred cooperatives, and sixteen hundred foreign importing organizations, which generally contribute $5,000 each per year. They benefit from the financial participation of American farmers in the form of voluntary dues in some states when the product is first delivered to the silo, slaughterhouse, or sugar factory.

This system of voluntary contributions is based on decentralized decision making which must be approved by a vote of the majority of producers of the commodity in question in the state concerned. Equaling between $1 and $2 for every $1,000 in product value, they can be repaid to reluctant farmers if necessary. The procedure for recuperating this money is complex, however, and the demands for reimbursement never represent more than 2 to 3 percent of the tax. This amount is paid to a state commission responsible for the product in question, which then divides it among its various uses: research and development, domestic promotion, and export development.

The Market Development Cooperators spent $35 million in 1977 of which $11 million came from USDA, $10 million from farmers and U.S. firms, and $14 million from foreign importers. The main products concerned were cotton ($9.4 million), fruits and vegetables (6.7), wheat and rice (6.3), soybeans (5.0), feed grains (3.3), meat and livestock (2.4), and poultry (1.2).

The promotion of American wheat exports, for example, is accomplished essentially by two groups, which for the last twenty years have undertaken considerable efforts in the field of information and propaganda. Western Wheat Associates and Great Plains Wheat, Inc., associations of farmers and industrial users of wheat, financed by parafiscal taxes on wheat, have a private status but work in close collaboration with the USDA.

Western Wheat Associates is supported by wheat producers in the western United States, and its activity is primarily oriented toward the promotion of American wheat in the Far East. This association has promotion offices in New Delhi, Tokyo, Seoul, Rizal (Philippines), Singapore, and Taipei. Great Plains Wheat, Inc., which represents wheat producers from the seven midwestern states (the Dakotas, Nebraska, Kansas, Oklahoma, Texas, and Colorado), concentrates its activities in Europe, Africa, the Mideast, and Latin America. Created in 1959, this association has the principal aim of stimulating American wheat sales through a variety of methods, such as technical assistance, market information, consumer promotion, and contacts between official representatives of client countries and U.S. government representatives. Great Plains

Wheat, Inc., is headquartered in Washington, where it maintains close relations with the federal government and grain merchants. It has four overseas regional offices: one in Rotterdam covering Europe and the Middle East; another in Caracas for Venezuela, Colombia, Central America, and the Caribbean; and a third in Rio de Janeiro for Brazil, Peru, Equador, Chile, and Bolivia. The fourth was created in 1978 in Casablanca to cover Africa. Furthermore, Great Plains Wheat is participating financially in the market development efforts of Western Wheat Associates. Great Plains and Western Wheat merged in 1979 under the name U.S. Wheat Associates, Inc.

Other professional organizations are particularly dynamic overseas. Mention should be made of the U.S. Feed Grains Council, an organization with offices in Paris and eleven other foreign cities: London, Hamburg, Rotterdam, Rome, Athens, Madrid, Warsaw, Singapore, Tokyo, Seoul, and Taipei. These offices disseminate information not only on the use of livestock feed but also on the new American animal husbandry techniques.

The American Soybeans Association, taking advantage of the great reputation of this nutrient, has been able to promote the use of soy meal in animal feed by using its seven foreign offices in Brussels, Hamburg, Madrid, Tokyo, Taipei, Vienna, and Mexico City.

The U.S. Meat Export Corporation is the latest of these export promotion organizations. This association, which was recognized as a market development cooperator by the USDA in 1976, is open to any group or individual concerned with the production, processing, and commercialization of beef, pork, or lamb. In 1980, it had only two foreign offices, one in Tokyo, covering all of Asia, and the other in London, covering Europe.

The development activities of these associations fall into two categories: (1) a service to the foreign importers and distributors to assist them in the use of American products by local consumers, either directly (conditions of the product's use, technical characteristics, and so on), or indirectly (encouraging the adoption of techniques that will lead to the use of American products—for example, the development of intensive cattle raising on feedlots, industrial husbandry of poultry to encourage the consumption of corn and soybeans, assistance in the creation of bakeries to promote American-type bread; and (2) promotional activities to encourage use of the product by foreign consumers themselves through the traditional means: publicity, in-store promotion, and the like. This type of promotion is reserved for value-added products, which, because of their characteristics, can easily be identified as American, such as Florida frozen orange juice or California almonds.

The American export drive has been highly effective: Food and agricultural exports reached $40 billion in 1980, compared with $6 billion in 1970.

The promotion of the American food and agricultural model is assured, as we have seen, by a massive and complex administrative apparatus. The activities of this complex, which achieves undeniably successful results, can be summarized in three points:

1. It is based on an efficient and controlled circulation of information at all levels. Data are gathered, processed, and disseminated methodically. The receipt of information by its intended recipient and its effective use are well controlled.

2. The diversity of sources of program financing (grants, short- to long-term credits, farmers, firms, and state contributions) is also characteristic of the American system. It partially frees the system from federal budget constraints, which are currently severe.

3. Decisions on the operation of programs and the observation of foreign markets are fairly centralized, particularly because of the key role given to the seventy agricultural attaché offices. Promotional activities, on the other hand, are basically decentralized, though coordinated.

The American model of agriculture is so built-in that it gives all the chances to private activities to develop with the help of U.S. government tools whenever they are judged useful. American farmers are therefore the center of a formidable complex which is believed, with high technology, to be one of the two pillars of the U.S. economy in the future.

Conclusion: Challenges and Prospects

THE WEIGHT OF THE U.S. AGRICULTURAL ECONOMY ON THE WORLD ECONOMIC BALANCE

Despite the extraordinary development of industry witnessed in the world over the past one hundred years, and of agriculture over the past fifty years, the fundamental question posed by Malthus became intensely relevant once again following the Soviet purchases of American grain in 1972: Was a new world food crisis in the offing for the end of this century?

As it turned out, after the 1972 crisis, the world food and agricultural balance was reestablished in three years because of the agricultural production reserves of the developed countries, in particular the United States, which since 1976 has confirmed and accentuated its dominant role. Can the United States maintain this position in the years ahead?

American authorities, in any case, believe it can. This attitude is based on the conviction that the agricultural sector in the United States has reached a threefold equilibrium. The policies implemented in 1933, 1961, 1973, 1977, and 1985 seem to guarantee a long-term balance between supply and demand. Furthermore, these policies have minimized the cost of government intervention to American consumers and taxpayers and guaranteed until 1982 satisfactory income for farmers and stable feed supplies for livestock breeders.

This success has been made possible only through profound structural changes, which have tended to favor the development of both the largest family farms and small subsistence farms. It is difficult, however, to formulate policies appropriate for both large and small farms, on the one hand, and for both livestock producers who want to buy cheap feed and grain and oilseed producers who want to export their products at the highest possible price, on the other. It is nonetheless noteworthy that all of these groups have made continuous progress in the United States.

Growth in American agriculture continues year in and year out. Production and export records are constantly being challenged. American agriculture, continually in search of new markets, responds immediately to each jump in

175

world demand. That 85 percent of supplemental feed grain exports throughout the world between 1973 and 1980 was supplied by the United States shows the enormous influence of this country on the development of livestock feed consumption.

The speed of the American producers' reaction results from a classic chain of events. The initial supplemental demand is first satisfied by stocks carried over to the end of the crop year. These final stocks have been generally available since the end of the 1972–75 crisis and since the creation of the farmers-owned reserve program introduced by the 1977 farm bill. A reduction in forecasted closing stocks then causes a rise in prices, which in turn encourages farmers to plant a greater portion of their land with the crops in most demand. Inversely, a reduction in demand or an excessive crop area for a given product causes an increase in quantities forecasted to be carried over to the following crop year, a decline in prices, and consequently a freeing up of acreage, which can be planted with a crop in greater demand on the world market.

This system does, however, present certain dangers, as seen in 1972–75, when the lack of sufficient reserves led to a tripling in grain prices, as well as in 1976–77 and 1983–84, when particularly abundant world harvests caused a doubling in U.S. grain stocks. A voluntary reduction in crop area, however, made it possible to limit the growth of these stocks and resume the increase in price the following year.

Some authors believe that the increase in production costs brought on by the energy crisis will put an end to this process. The importance of energy costs in the American agricultural system, however, should be nuanced. Energy costs represent less than 10 percent of production costs, which themselves represent a little less than 90 percent of farm sales. It was necessary to provide these figures to dispel the impression that American agriculture is wasteful of energy. They show, as will be increasingly recognized, that throughout the world agriculture is one of the activities that will most effectively withstand the effects of oil price increases.

Other observers are alarmed by the erosion of American soil. Yet, though it is true that some areas may be threatened by negligence in agricultural methods, the vast majority of American farmers are aware of this danger and the correct procedures to avoid it, as are the research centers and government services.

Countries that have been able to maintain the technical and economic level of their farming sector will find themselves endowed with renewable wealth, which countries having sacrificed their agricultural sector to urban growth will sorely lack. But this raises a question. Will the strength of American agriculture endow the United States with a frightening power over the rest of the world?

DOES GREEN POWER EXIST?

Assured of maintaining its leadership role in agricultural commodity exports for at least the next twenty years, the United States could be tempted to use this green power to further its international objectives. A number of other countries could also be seduced by the strategic implications of food problems during the remainder of the twentieth century.

Three approaches considered incompatible with free market principles could be taken by this country to impose its political will. In order of declining restrictiveness these are the embargo, the creation of a cartel, and product agreements or negotiations within the international organizations. It may appear paradoxical that the order of effectiveness of these economic weapons runs inversely.

The embargo ordered by President Carter on January 4, 1980, on the supply of grain and other products essential for the food and agricultural economy of the Soviet Union would appear to demonstrate that food power exists, since it was used in a specific policy decision. The results of this experience, however, show that this conclusion must be qualified. To avoid alienating American farm voters and openly repudiating an agreement signed by the United States government,[1] President Carter did not order an immediate and total embargo. The 8 Mt of grain which the United States had agreed to deliver to the Soviet Union under the five-year agreement signed October 20, 1975, were exempted from the embargo, leaving 3.8 Mt available for delivery after January 4, 1980.

With the embargo thus emasculated, the temptation was great for countries less subject to American influence, or less concerned about the Soviet presence in Afghanistan, such as Argentina, to profit from the windfall of this black market outlet, offering considerably higher prices than those available on the world market. Thus Soviet imports for the period from July 1979 to June 1980 reached 31 Mt of grain, although the United States had intended for them to decrease to 23 Mt (the initial projection had been 35 Mt). The goal of drastically reducing Soviet meat production, therefore, had proven unattainable. By drawing on their stocks, the Soviets had been able to maintain grain reserved for livestock feed at approximately 120 Mt (see Appendix 19).

The results were therefore mediocre, despite the size of the Soviet grain deficit in 1979–80: 45 Mt. Nevertheless, the expansion of livestock production targeted for the USSR in the five-year plan seems to have been blocked for some years, and the relative facility with which the American measure was accepted by U.S. farmers and allied countries showed that the social cost of such a measure was still bearable. The budgetary cost of the compensation accorded to American farmers did not surpass $1 billion once the resulting stocks had been sold off on the market.

The qualified results of the agricultural embargo must be placed in the context of the world food and agricultural equilibrium. On the one hand, the United States has clearly become the only country possessing sufficient power in world corn and soybean markets to consider resorting to an embargo. On the other hand, a number of potential targets for the food weapon can be identified.

The largest deficit exists in Western Europe, particularly the EEC, which imports 80 percent of its vegetable protein and about one-fourth of its energy foods, primarily based on grain or grain substitutes, such as cassava, corn gluten feed, and even soybeans for the nonprotein part of the meal. With 120 million inhabitants and practically no arable land, Japan is clearly vulnerable. Its food deficit exceeds $12 billion and should continue growing. The Soviet Union and Eastern Europe constitute a third zone with fragile food supplies. With the evolution of their agricultural economy toward large-scale meat production based on prepared feed, the further their five-year plan progresses, the greater their dependence becomes. Given their strategic role, these countries appear to be the most likely targets for an embargo. Yet, just when the socialist countries are adopting the American agricultural model, that is, giant-scale livestock operations, with all the technology and raw material imports that entails, was it conceivable that the United States would prolong indefinitely a policy that would become increasingly tenuous over time? The Reagan administration responded in the negative, lifting the embargo in April 1981, just when the announcement of another bad Soviet harvest increased the chance of success of the decision taken by his predecessor, and signing on August 25, 1983, a new five-year grain agreement raising the minimum annual Soviet purchase from 6 to 9 million tons and the possible purchase without negotiation from 8 to 12 million tons.

China could at some point also have become a target of the food weapon if its imports, which exceeded 15 Mt of grain until 1983, had continued to grow. But the new agricultural policy implemented in 1981 seems to have reversed the former trend. The OPEC countries, on the other hand, are less vulnerable. The small populations of the countries in this area could be supplied with relative ease by nearly any competitor of the United States. Indeed, food imports by the Middle Eastern countries are less than $20 billion and are highly diversified.

The embargo is clearly a difficult measure, and it is limited both temporally and geographically. Yet it is not the only possible measure. The United States might also be tempted by what could be called a cartel policy. This would consist of responding to OPEC with an equivalent strategy, that is, an agreement among grain- and soybean-producing countries to increase the price of these basic commodities. This is the idea behind the famous slogan "a barrel of oil for a bushel of wheat,"[2] started by the National Farm Organization, an American agricultural association.

A number of meetings initiated by Robert Bergland, agriculture secretary under the Carter administration, with his colleagues from Canada, Australia, and occasionally Argentina and Europe, may have led to the impression that the establishment of such a grain cartel was under way. The meetings held in Saskatoon (Canada, May 1979), Buenos Aires (Argentina, October 1979), Washington (United States, January 1980), and Brussels (EEC, June 1980), however, appeared at the most to allow concerted policies among the five major grain-exporting countries to avoid excessive price decreases on the world market. These meetings have since become institutionalized and hence somewhat trite. Above all, they have allowed the Americans, who through their agricultural attachés possess the greatest information network, to share their analysis of the market situation and to explain their stock management measures. But as yet, they have not led to specific and coordinated measures to increase prices.

In reality, OPEC, the intended target, would not be overly concerned. Since they purchase primarily processed food products, as opposed to basic commodities, the oil-producing countries would be much less affected by this increase in prices than Western Europe and Japan would be. As far as the countries with collective agriculture are concerned, they have not been dependent on Western grain long enough to prevent them from imposing import restrictions if prices increased drastically.

Soybeans are a simpler target for cartel action. An agreement between the United States and Brazil, possibly including Argentina and Paraguay, would suffice to control nearly all exports. An export tax could then be imposed, sending the price of vegetable protein skyward. In that case, the targets affected would be primarily the more advanced agricultural operations in the countries allied with the United States: Europe and Japan (see Appendix 16). Such a measure would cause a tremendous decrease in soybean consumption, particularly in Europe (see Chapter 2, Table 2.3), an area where consumption is extremely high, given the highly favorable price of soybeans in comparison with that of corn. So, though it would be the easiest to implement, a soybean cartel would be the least advantageous from a strategic standpoint.

Any cartel policy is difficult to carry out, first, because within the allied countries it is difficult to maintain the delicate balance between the interests of producers and livestock breeders. Furthermore, as in any market situation, equilibrium is often unstable, and the temptation is great for members to back out of the agreement. Indeed, as in the case of the embargo, price increases encourage outside competition. Moreover, although the members of the potential alliance, the United States, Canada, Australia, Argentina, and the EEC, account for nearly 90 percent of world grain exports, they contribute only 42 percent to total world grain production. The other countries would have to increase their national productions by only 8 percent to compensate for export reductions caused by a voluntary increase in world prices.

This argument, however, is not entirely convincing. The additional 8 percent would, by necessity, be produced at a higher marginal cost than that enjoyed by the cartel. It is therefore relatively easy, as under the price limit theory, to make market entry very costly from one year to the next. This is clearly one of the essential differences between the food weapon and the oil weapon. Although the OPEC countries produce most of the world's oil and consume only a small fraction, the countries accounting for 90 percent of grain exports produce less than half of the world's grain and consume more than two-thirds of their production. This is why in 1974, when the prices of wheat and oil simultaneously staged dramatic increases, American farm income from grain and soybeans increased only by $8 billion over that in 1973, while OPEC's oil income increased by $60 billion.[3]

As a source of energy, agricultural products cannot compete effectively with oil. Oil is a natural source of energy, prepared by eons of biological and geological interaction. Agricultural products, on the other hand, are produced by men attempting to capture as much solar energy each year as possible by improving the synthesis of chlorophyll. As a form of energy, biomass offers little promise. The use of agricultural products as an alternative source of energy, replacing oil, is conceivable only for short periods of time when energy prices are extremely high. Any nonfood use of agricultural products such as sugar or grain can lead only to rapid price increases for these products. Competition between food and energy applications would lead to dramatic worldwide inflation. Economically viable biomass consists of nonfood by-products of agricultural or forest production. In the long run, it is more advantageous to export sugar or wheat than to convert it into alcohol.

It is understandable, under these circumstances, why U.S. foreign agricultural policy is limited to efforts in favor of lowering tariff and nontariff barriers to trade in food and agricultural products. Aside from a few bilateral negotiations with the EEC, Japan, Brazil, and other countries, the U.S. offensive has always been deployed within the framework of the GATT multilateral trade negotiations.

The results of the Tokyo Round negotiations[4] do not allow the conclusion that American pressure for more liberalized agricultural trade will subside. Granted, the United States conceded nothing in the area of dairy product protectionism (see Chapter 4). It also obtained a code of conduct on the allocation of export subsidies, including those for agricultural products, and it secured an opening for its high-quality hilton beef in the EEC market. But the primary result of these negotiations for the United States was to maintain an open door into the EEC for soybean exports. The "consolidation" of no duty and no levy on this product, obtained during the Dillon Round in 1963, represented a value of $2.5 billion in soybean and oil cake exports to the EEC in 1980, whereas the initial concession seventeen years earlier applied to only $700 million. Moreover, the Tokyo Round extended this consolidation of soybeans to Japan and opened the possibility of freely exporting to this country $100 million worth of beef. By 1997, Japan could well be a massive importer of American beef.

Continuous efforts to develop overseas agricultural market access, therefore, turn out to be more effective than brusque, unproductive embargoes, pressure by a cartel on prices, or international agreements. Agreements would enable the United States to share the burden of maintaining stocks with other exporting and importing countries, but they are not supported by American farmers.

Indeed, "even assuming that Europe resumes agricultural expansion and that the Don plains have a mild summer, the tide of Third World demand is rising inexorably. American farmers face a promising future."[5] This opinion is largely shared in the Midwest because American farmers know from experience that the food weapon does indeed exist. Yet it is a delicate weapon;[6] it must be rattled from time to time but is best never used. It loses its edge quickly. The best means of using it is to continue producing in response to world demand and always be the fastest to meet unforeseen needs created by climate aberrations. This is the strength of the American model.

THE REAL STAKES: IS THE AMERICAN MODEL THE ONLY MEANS TO WORLD AGRICULTURAL DEVELOPMENT?

It is misleading to contrast the free trade policies adopted by American agriculture with the more or less admitted inclination of European policy toward protectionism. This perspective is too simplistic. American policy can be either protectionist or oriented toward free trade depending on the products involved and the amount of foreign competition.

American agricultural policy is distinguished from other policies by its systematic and constant use of a complex set of economic tools. This sets it apart from policies elsewhere that could be termed improvised, ad hoc, or naive. But does this mean that the American model is the sole standard for world agricultural development? To answer this question, it is first necessary to define the concept of "American model" more clearly. Two definitions are possible, one technological, the other economic.

The first of these two definitions was analyzed in Chapter 6. It can be summarized, somewhat in caricature, as a world in which the United States produces practically all the feed grains and oilseeds it supplies to countries requiring them for livestock feed. The second definition is an extension of the first. It upholds the idea of relatively free and competitive markets, lowered customs tariffs, and commodity prices that freely reflect international trade conditions.

In this context it seems clear, and a number of experiences have demonstrated that there are other approaches to agricultural development. The Eastern bloc countries thought they could substitute a centrally planned system of supplies to the state for the free market—an incentive-based agricultural system. The unexpected result of the Soviet policy was a concentration of about 30 percent of agricultural production within individual plots and family gardens occupying less than 4 percent of total agricultural land area. This distortion of resource allocation represents a considerable diseconomy severely hindering the agricultural development of these countries.

Some developing countries have often confused agricultural development with the construction of push-button agricultural operations. Neglecting the agricultural masses, which constitute more than 60 percent of their population, countries of completely opposite political orientations have invested considerable sums in giant capitalist or state farms, resulting only in an acceleration of rural migration toward the urban outskirts.

The cases of China, India, and Brazil, however, should be examined separately because they have displayed a concern for basing development on the potential of their agricultural populations. Forced collectivism, which characterized the Chinese Cultural Revolution, interrupted the balanced process of agricultural development that had been implemented after the "long march." The return in 1982 to a policy favoring human-scale farming operations produced quick results—grain production jumped from 268 Mt in 1981 to 342 in 1983.

The case of India is much more promising. The extraordinary economic activity among the 80 percent of the population living in rural areas has been fruitful. The world's largest recipient of food aid in 1967, India has been self-sufficient in grain and sugar since 1975 (its grain production increased to 110 Mt in 1970, to 140 Mt in 1980, and should reach 180 Mt in 1990) and possesses the world's largest dairy livestock population.

Brazil, finally, has achieved astonishing agricultural growth allowing it to offset a large portion of its oil imports with exports of coffee and soy meal. Development in the southern portion of the country, however, is accompanied by continued poverty in the Nordeste.

The EEC agricultural policy has been partially successful in promoting an agricultural sector providing self-sufficiency in food for its 330 million inhabitants. Although the goal is commendable, the results have been only partially achieved, despite the considerable resources applied. The percentage of food and agricultural imports into the EEC offset by exports is only about 30 percent, but the rate of overall self-sufficiency is nearly 85 percent. The system used by the common agricultural policy was, until recently, based on rigid, annually set prices. This system resulted in the production of surpluses in dairy products, barley, and occasionally meat, which were difficult to sell, but considerable deficits occurred in other feed grains, vegetable protein, fruits, and vegetables. Nonetheless, it allowed intensive grain production exceeding 150 Mt in 1984 to be maintained on 27 million hectares.

The important agricultural countries of the Commonwealth constitute a separate entity. They have followed U.S. policy with respect to the orientation of agricultural supply based on world prices but have resorted to highly active government intervention to assure exports of grain, dairy products, and beef and lamb. The Marketing Boards in Canada, Australia, and New Zealand are examples of Anglo-Saxon pragmatism in sectors where swift commercial reaction is essential.

The relative success of some of these "other options" must not, however, create any illusions. It is one thing to achieve adequate agricultural production, but quite another to challenge American agricultural dominance in a world market increasingly subject to the laws of competition. For now, there is no country, nor even any group of countries, capable of meeting that challenge.

APPENDIXES

Comparison of the developments in agricultural production and the population

100=1961–65	FAO per Capita Agricultural Production Index		
	1966–70	1971–75	1976–80
World	103.2	105.8	108.9
Africa	99.8	97.3	91.6
North and Central America	100.4	103.0	108.5
South America	100.0	100.7	107.9
Asia	103.0	106.1	112.5
Europe	109.2	117.8	124.7
Oceania	105.0	105.6	110.9
USSR	112.4	119.0	112.8
Developed countries with a market economy	105.2	110.2	116.6
Developing countries with a market economy	100.2	101.2	104.9
Countries with a centrally planned economy	108.2	114.1	118.3

Source: FAO, 1971, 1976, and 1981 Production Yearbook, Table 7.
Note: A number higher than 100 signifies an improvement in the individual food situation.

Economically speaking, it is indispensable to make a clear distinction between three concepts: yield increases (average productivity), marginal productivity, and increasing costs.

If agricultural production were carried out in a world in which inputs could be used without limitations on quantity, yields would be constant. Two conclusions can be drawn from this. The first is that a doubling of all inputs (land, labor, capital, and so on) would lead to a doubling of output. The second conclusion is that the proportion in which these inputs are used would not change when doubled as long as the prices for inputs do not change. If the relative costs of inputs were to be modified, the farmer would adjust to this new situation by changing technology. In this utopian world, the concept of the return on a given input, such as land, has no real meaning because it is only when combined with the other inputs, at a given price level, that it is truly effective. A good measurement standard in this case is the marginal productivity of this input—how much will production be increased, at the margin, when an additional unit of input is used?

In the real world, inputs are not infinitely available. Some are quite limited, and it is traditional to consider land as an invariable factor in agriculture. Consequently, the producer has a given arable land area, and depending on prices expected for his crops and the cost of other inputs, he determines the quantity of production inputs he will use. This monitored freedom has an immediate consequence: if he wanted to produce more than before he would have to use more variable inputs (labor, capital) on the same land area. Just as enjoyment of a nightclub increases the more crowded it becomes—up to the point that the dance floor becomes unbearably crowded—production increases with the addition of variable inputs until the land area, which is constant, becomes gorged with workers, machinery, and fertilizer. Yield increases, measured as the number of units produced per acre, begin to diminish. Furthermore, more intensive use of variable inputs is generally accompanied by an increase in their cost because greater demand increases the short-term price. The farmer is then faced with both diminishing yield growth and increasing costs. He stops producing when the marginal profit is zero. In this situation the concept of yield, or average productivity, can be clearly understood because it is measured in terms of a constant factor, land. When this constant factor begins to vary because new arable land has been made available, it is absurd to make yield comparisons between the old situation and the new one. Entirely new technology will be applied, which will not reach its optimal productivity until the land factor once again becomes constant. In the meantime, it is perfectly possible that the new technology will lead both to increasing costs, because it consumes more inputs, and to increasing yields, with the latter phenomenon being the determining factor justifying agricultural specialization.

APPENDIX 3. The elasticity of food demand as compared to individual income, 1973

Gale Johnson Hypothesis				FAO-OECD Hypothesis		
Per Capita GNP ($)	r	Region	r	Region	r_1	r_2
165	0.9	North America	0.16	North America	−0.01	0.04
260	0.7	Oceania	0.10	Oceania	0.02	0.05
491	0.6	EEC	0.47	Western Europe	0.07	0.29
910	0.6	Denmark	0.19	Others	0.14	0.47
1,285	0.5	Sweden	0.20	Total of developed countries	0.08	0.24
1,440	0.2	United Kingdom	0.24	Africa	0.29	0.95
2,190	0.16	Latin America	0.40	Latin America	0.20	0.42
		Near East/Africa	0.68	The Near East	0.15	0.52
		Asia and the Far East	0.89	Asia and the Far East	0.34	0.83
		Japan	0.58	Total of developing countries	0.22	0.58

Sources: Gale Johnson, *World Agriculture in Disarray* (London: Macmillan/Saint Martin Press, 1973); OECD, *Etude des tendances de l'offre et de la demande des principaux produits agricoles* (Paris: OECD, 1976).
r = elasticity for calories of animal origin.
r_1 = elasticity for total food calories.
r_2 = elasticity for calories of animal origin.

APPENDIX 4. The productivity and size of farms in the United States

Year	Productivity Index 100 = 1967	Changes per Decade (%)	Average Land per Farm (acres)	Changes per Decade (%)
1910	53	−	138	−
1920	54	+1.8	147	+6.2
1930	53	−1.8	151	+2.7
1940	62	+17.0	167	+10.6
1950	73	+14.5	213	+27.5
1960	92	+26.0	297	+39.4
1970	101	+20.7	373	+25.6
1975	113	+23.8	387	+7.6

Source: B. Gardner and R. Pope, "How Is Scale and Structure Determined in Agriculture," *American Journal of Agricultural Economics,* May 1978, pp. 295–308.

APPENDIX 5. **The Silo Plan**

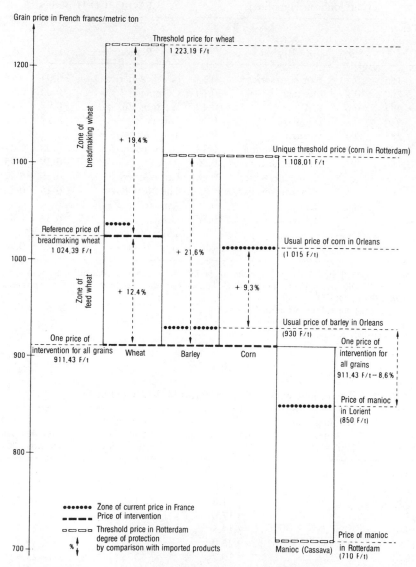

Source: AGPB (the French Association of Wheat Growers), Congrès de Toulouse, June 10–11, 1980

APPENDIX 6. The largest agricultural countries in the world (in cropland and in grain production)

Classification by Land Area	Countries or Groups of Countries	Cropland and Permanent Crops in 1980 (millions of hectares)	Grain Production in 1981 (millions of tons)	Classification by Production
–	World	1,452	1,664	–
–	Developing countries	781	811	–
–	European countries with a centralized economy	278	247	–
–	Developed countries with a market economy	394	605	–
1.	USSR	232	167	3d
2.	United States	191	334	1st
3.	India	169	150	4th
4.	China	99	286	2d
	Total of countries 1–4[a]	47.6%	56.3%	–
–	12-member EEC	79	136	–
–	9-member EEC	55	118	–
5.	Brazil	62	32	8th
6.	Australia	45	23	12th
7.	Canada	44	50	5th
8.	Argentina	35	31	9th
9.	Nigeria	30	10	27th
10.	Turkey	28	25	11th
	Total of countries 1–10[a]	64.4%	66.6%	–
11.	Mexico	23	26	10th
12.	Spain	21	12	25th
13.	Pakistan	20	18	19th
14.	Indonesia	20	37	7th
15.	France	19	45	6th
16.	Thailand	18	23	13th
17.	Iran	16	9	28th
18.	Poland	15	20	16th
19.	Ethiopia	14	4	–
20.	South Africa	14	18	20th
	Total of countries 1–20[a]	76.8%	79.2%	–

Source: FAO, Production Yearbooks, 1977 and 1981.

Note: A certain number of countries whose cropland is less than 12 million hectares are classified by their grain production in the following manner: Federal Republic of Germany ranks fourteenth, Romania fifteenth, Great Britain seventeenth, Italy eighteenth, Yugoslavia twenty-first, Japan twenty-second, Viet Nam twenty-third, Greece twenty-fourth, and Philippines twenty-sixth.

[a]In percentage of world land area and production.

APPENDIX 7. The agrofood billionaires of foreign trade (in billions of current U.S. dollars)

Rank in 1982	Country	Value in 1982	Value in 1977	Rank in 1977
A. *The major exporters*				
1.	United States	38.2	24.8	1
2.	France	15.8	9.3	3
3.	Netherlands	15.1	10.0	2
4.	Federal Republic of Germany	10.6	6.2	5
5.	Australia	8.7	5.9	6
6.	Brazil	8.0	7.2	4
7.	Canada	8.0	4.3	9
8.	Great Britain	7.4	4.3	8
9.	Belgium	6.1	4.0	10
10.	Italy	5.6	3.3	11
11.	Argentina	5.0	4.3	7
12.	Denmark	5.0	3.1	12
13.	Thailand	3.9	2.2	16
14.	Cuba	3.8	1.7	23
15.	New Zealand	3.5	2.1	17
16.	China	3.3	2.2	15
17.	Spain	3.0	2.1	18
18.	Malaysia	3.0	2.5	14
19.	USSR	2.8	2.7	13
20.	Turkey	2.5	1.1	29
21.	Ireland	2.4	1.8	22
22.	India	2.3	1.5	25
23.	Hungary	2.2	1.8	20
24.	Colombia	2.1	1.9	19
25.	South Africa	1.9	1.4	27
26.	Singapore	1.7	0.9	32
27.	Philippines	1.5	1.6	24
28.	Indonesia	1.5	1.8	21
29.	Ivory Coast	1.5	1.5	26
30.	Greece	1.3	1.0	31
31.	Mexico	1.3	1.3	28
32.	Bulgaria	1.3	1.0	30
B. *The major importers*				
1.	Federal Republic of Germany	23.1	18.0	1
2.	USSR	19.4	9.7	6
3.	United States	16.9	14.1	2
4.	Japan	16.2	12.7	3
5.	Great Britain	14.0	12.3	4
6.	Italy	13.3	9.5	7
7.	France	12.9	10.8	5
8.	Netherlands	10.3	7.5	8

Rank in 1982	Country	Value in 1982	Value in 1977	Rank in 1977
9.	China	7.4	3.7	10
10.	Belgium	7.4	5.6	9
11.	Saudi Arabia	5.1	1.5	23
12.	Canada	4.4	3.5	11
13.	Spain	3.8	3.2	12
14.	Hong Kong	3.4	2.1	14
15.	Egypt	3.2	1.6	22
16.	South Korea	3.0	1.6	20
17.	Switzerland	2.8	2.2	13
18.	Algeria	2.4	1.3	28
19.	Nigeria	2.3	1.4	25
20.	Singapore	2.2	1.2	29
21.	Iran	2.2	1.9	16
22.	Poland	2.2	2.0	15
23.	German Democratic Republic	2.2	1.9	17
24.	Denmark	2.1	1.7	19
25.	Iraq	1.9	0.8	37
26.	Sweden	1.9	1.8	18
27.	Czechoslovakia	1.9	1.6	21
28.	Mexico	1.8	0.8	34
29.	Venezuela	1.8	1.3	27
30.	Brazil	1.8	0.9	32
31.	India	1.5	1.4	24
32.	Portugal	1.5	1.1	30
33.	Austria	1.5	1.3	26

C. *Main agricultural trade surplus countries*

1.	United States	21.4	10.7	1
2.	Australia	7.6	5.2	3
3.	Brazil	6.2	6.2	2
4.	Netherlands	4.8	2.7	5

(*continued*)

Rank in 1982	Country	Value in 1982	Value in 1977	Rank in 1977
5.	Argentina	4.6	4.1	4
6.	Canada	3.6	0.8	17
7.	Thailand	3.4	1.9	7
8.	New Zealand	3.1	1.9	6
9.	Denmark	2.9	1.4	10
10.	France	2.8	−1.0	−
11.	Cuba	2.7	1.1	13
12.	Turkey	2.2	1.0	15
13.	Malaysia	1.6	1.7	8
14.	Colombia	1.5	1.5	9
15.	Hungary	1.5	0.6	−
16.	South Africa	1.2	1.0	16
17.	Ireland	1.2	1.0	14
18.	Ivory Coast	1.0	1.3	11
19.	India	0.8	0.5	−
20.	Philippines	0.8	1.2	12
21.	Bulgaria	0.8	0.6	−

D. *Main agricultural trade deficit countries*

Rank in 1982	Country	Value in 1982	Value in 1977	Rank in 1977
1.	USSR	−16.6	−6.6	4
2.	Japan	−15.4	−11.9	1
3.	Federal Republic of Germany	−12.5	−11.8	2
4.	Italy	−7.6	−6.2	5
5.	Great Britain	−8.0	−6.6	3
6.	Saudi Arabia	−5.0	−1.2	15
7.	China	−4.1	−1.4	11
8.	Egypt	−2.5	−0.8	20
9.	South Korea	−2.5	−1.1	16
10.	Hong Kong	−2.4	−1.7	7
11.	Algeria	−2.3	−1.0	18
12.	Iran	−2.1	−1.8	6
13.	Nigeria	−2.0	−0.7	22
14.	Iraq	−1.9	−0.8	21
15.	Switzerland	−1.8	−1.5	9
16.	Venezuela	−1.8	−1.3	13
17.	German Democratic Republic	−1.6	−1.4	10
18.	Poland	−1.4	−1.0	19
19.	Czechoslovakia	−1.4	−1.2	14
20.	Belgium	−1.3	−1.6	8
21.	Sweden	−1.1	−1.3	12
22.	Portugal	−1.1	−0.7	23
23.	Spain	−0.8	−1.1	17

Source: FAO, *Trade Yearbook,* 1977 and 1982.

APPENDIX 8. **Shares in the 1977 market for seven groups of products representing 53 percent of world exports (153 billion dollars) (in percentage of world exports of each group)**

Grains $22.0 billion		Coffee-Cocoa-Tea $18.2 billion		Oilseeds, Beans, and Meals $10.5 billion		Meat $9.9 billion		Sugar $7.1 billion		Milk and Dairy Products $7.0 billion		Vegetable Oils $5.2 billion	
United States	38.6	Brazil	15.8	United States	54.6	Netherlands	14.9	Cuba	21.8	Netherlands	21.3	Malaysia	14.3
Canada	10.0	Colombia	8.3	Brazil	18.0	United States	7.6	Australia	10.3	France	17.2	United States	13.2
France	8.6	Ivory Coast	7.3	Argentina	4.8	New Zealand	7.3	Philippines	7.6	Federal Republic of Germany	16.2	Philippines	8.1
Argentina	7.7	Ghana	4.4	Canada	3.0	Denmark	6.9	France	7.5	Denmark	7.8	Brazil	7.7
Australia	6.8	Kenya	3.7	Netherlands	1.8	France	6.2	Brazil	6.5	New Zealand	6.2	Federal Republic of Germany	7.5
Thailand	3.6	Indonesia	3.5	Federal Republic of Germany	1.8	Ireland	6.0	Thailand	5.1	Belgium	5.3	Argentina	5.5
USSR	3.1	El Salvador	3.3	India	1.8	Federal Republic of Germany	5.5	Federal Republic of Germany	3.3	Ireland	4.8	Netherlands	4.5
Netherlands	2.7	India	3.2			Argentina	4.8	South Africa	3.2	Australia	3.1	Spain	4.2
		Netherlands	3.0			Belgium	4.6	Dominican Republic	3.1	Switzerland	3.0	Senegal	3.5
		Nigeria	3.0			Hungary	4.3			Great Britain	2.2	France	3.4
		United States	2.7			Great Britain	4.2			Italy	1.4		
		Guatemala	2.7							Austria	1.4		
		Mexico	2.5							Canada	1.3		
		Ceylon	2.4										
		Cameroun	2.1										
		Uganda	1.7										

Source: FAO, *Trade Yearbook, 1977.*

Note: Country figures are expressed as percentage of world exports. Columns do not total 100 percent due to the omission of some countries with minor output.

193

APPENDIX 9. Net farm income and off-farm income of farmers in the United States in 1962, 1972, and 1982 (global income of each class in millions of current dollars)

Class of Sales (in current dollars)	Year	Net Farm Income	Off-farm Income	Total Income	Percent Net Farm Total
Class IC	1962	NS	NS	NS	NS
(sales greater than $500,000)	1972	3,330	NS	3,330	100.0
	1982	14,306	672	14,978	95.5
Class IB	1962	878	NS	NS	100.0
(sales between $100,000 and	1972	4,659	609	5,268	88.4
$499,999)	1982	8,341	3,254	11,595	71.9
Class IA	1962	1,518	390	1,908	79.6
(sales between $40,000 and	1972	4,589	895	5,484	83.7
$99,999)	1982	2,174	4,166	6,340	34.2
Class II	1962	2,100	481	2,581	81.4
(sales between $20,000 and	1972	3,445	1,298	6,743	72.6
$39,999)	1982	138	3,522	3,660	3.8
Class III	1962	2,592	791	3,383	76.6
(sales between $10,000 and	1972	1,929	1,912	3,841	50.2
$19,999)	1982	−205	4,839	4,634	−4.4
Class IV	1962	1,909	1,261	3,170	60.2
(sales between $5,000 and	1972	802	2,506	3,308	24.2
$9,999)	1982	−292	6,345	6,053	−4.8
Class V	1962	1,022	1,980	3,002	34.0
(sales between $2,500 and	1972	855	2,868	3,723	23.0
$4,999)	1982	−328	6,344	6,016	−5.5
Class VI	1962	1,425	5,701	7,126	20.0
(sales less than $2,500)	1972	−116	11,177	11,061	−1.1
	1982	−231	10,289	10,058	−2.3

Source: USDA, Economic Research Service, Economic Indicators of the Farm Sector, 1983.

APPENDIX 10. Development across thirty years of the distribution of American farms for each sales class, 1949–1979

Economic Classes of Farms	Number of Farms (in thousands)				Distribution			
	1949	1959	1969	1979	1949	1959	1969	1979
All farms	5,247	4,097	3,000	2,430	100	100	100	100
Commercial farms	2,180	2,175	1,624	1,862	41.5	53.1	54.1	76.6
Class I: $40,000 or more	48	106	202	638	0.9	2.6	6.7	26.2
Class II: $20,000 to $39,999	107	219	304	283	2.0	5.3	10.1	11.7
Class III: $10,000 to $19,999	342	503	369	289	6.5	12.3	12.3	11.9
Class IV: $5,000 to $9,999	739	693	381	327	14.1	16.9	12.7	13.5
Class V: $2,500 to $4,999	944	654	368	325	18.0	16.0	12.2	13.4
Noncommercial farms with sales of less than $2,500 (Class VI)	3,067	1,922	1,376	568	58.5	46.9	45.9	23.4

Sources: USDA, Economic Research Service, *The Expanding and the Contracting Sectors of American Agriculture,* Agriculture Economic Report 74, 1965; USDA, Economic Research Service, *Economic Indicators of the Farm Sector,* 1983.

195

APPENDIX 11. The United States food industry in 1981

	Sales (in billions of dollars)[a]	Jobs (in thousands)	Annual Growth Rate of Jobs over 9 Years (%)	Number of Establishments
Slaughter and processing of meat and poultry	66.3	312	−1.4	4,534
Dairy products	36.8	143	−3.0	3,731
Processing of grains:	45.0	b	b	b
Baked products	13.8	169	−1.5	3,062
Breakfast cereals	3.9	48	+1.8	324
Sugar, sweets, and chocolates	15.0	b	b	b
Canned and frozen goods:	b	b	b	b
Canned fruits and vegetables	13.3	108	−1.1	1,103
Frozen products	9.7	81	0.0	654
Oils and oil meal	20.1	b	b	b
Alcoholic beverages:	17.3	72	−1.0	504
Beer	10.4	46	−1.3	131
Spirits	3.8	15	⎫ −0.5	104
Wines and brandies	3.0	11	⎬	269
Soft drinks	17.5	119	⎰ −0.5	2,192
Coffee	5.9	b	b	b

Source: U.S. Department of Commerce, *Industrial Outlook,* 1978 and 1982.
[a]Sales at the level of production prices.
[b]Not available.

APPENDIX 12. Contract farming and corporate farming

Percentage of output	Under Contract 1960	Under Contract 1970	Incorporated 1960	Incorporated 1970	Total 1960	Total 1970
Wheat and rice	1.0	2.0	–	–	1.0	2.0
Fresh vegetables	20.0	21.0	25.0	30.0	45.0	51.0
Vegetables for canned goods	67.0	85.0	8.0	10.0	75.0	95.0
Beans and dried peas	35.0	31.0	1.0	1.0	36.0	32.0
Potatoes	45.0	45.0	30.0	25.0	75.0	70.0
Citrus fruits	60.0	55.0	20.0	30.0	80.0	85.0
Other fruits	20.0	20.0	15.0	20.0	35.0	40.0
Sugar beets	38.0	38.0	62.0	62.0	100.0	100.0
Sugarcane	40.0	40.0	60.0	60.0	100.0	100.0
Cotton	5.0	11.0	3.0	1.0	8.0	12.0
Tobacco	2.0	2.0	2.0	2.0	4.0	4.0
Oilseeds	1.0	1.0	0.4	0.5	1.4	1.5
Seeds	80.0	80.0	0.3	0.5	80.3	80.5
Total vegetable outputs	8.6	9.5	4.3	4.8	12.9	14.3
Feedlot cattle	10.0	18.0	3.0	4.0	13.0	22.0
Sheep	2.0	7.0	2.0	3.0	4.0	10.0
Pigs	0.7	1.0	0.7	1.0	1.4	2.0
Milk for drinking	95.0	95.0	3.0	3.0	98.0	98.0
Milk for processing	25.0	25.0	2.0	1.0	27.0	26.0
Eggs	5.0	10.0	10.0	20.0	15.0	30.0
Broilers	93.0	90.0	5.0	7.0	98.0	97.0
Turkeys	30.0	42.0	4.0	12.0	34.0	54.0
Total animal outputs	27.2	31.4	3.2	4.8	30.4	36.2
General total	15.1	17.2	3.9	4.8	19.0	22.0

Source: Ronald Mighell and William Hoofnagle, *Contract Production and Vertical Integration in Farming, 1960 and 1970,* Economic Research and Statistical Service, no. 474 (Washington, D.C.: USDA, 1972).

APPENDIX 13. The importance of farm cooperatives, 1980 (billions of dollars)

	Number of Cooperatives Handling	Gross Sales of Cooperatives	Net Purchase or Sales to Members	Global Value of the Crop or the Supply	Cooperative Market Share (%)
1. Marketing Cooperatives					
Grain, soybeans	2,434	28.5	17.8	44.8	40
Milk	487	15.1	13.7	16.6	83
Beef	449	6.1	5.6	31.5	18
Fruits and vege-tables	400	5.8	4.2	13.6	31
Cotton	470	2.2	1.9	4.5	22
Poultry	74	1.2	1.1	4.4	25
Rice	60	1.0	1.0	1.5	67
Total 1	4,350[c]	63.7[d]	48.9[d]	139.5[d]	35
2. Supply Cooperatives					
Petroleum products	2,921	9.7	5.1	[a]	25[b]
Fertilizer	3,802	6.5	3.5	9.9	35
Animal feeds	3,613	5.3	3.5	18.6	19
Pesticides	3,697	1.8	1.1	[a]	33[b]
Seeds	3,620	0.7	0.5	3.4	14
Farm machinery and equipment	1,827	0.7	0.4	[a]	10[b]
Building materials	2,107	0.6	0.5	[a]	[a]
Total 2	5,012[c]	27.7[d]	16.1[d]	130.5[d]	13

Sources: USDA, Agricultural Statistics, 1982, Table 625, "Crop Value"; Table 658, "Farm Production Expenses"; Table 689, "Farmers' Cooperatives, Gross Business and Net Business"; A. Seguin, "Dossier de la Cooperation Agricole aux Etats-Unis," Agriculture Cooperation (Paris), July–August 1979.

[a]Unavailable.
[b]1978 evaluations.
[c]Total is less than sum because many coops do multiple businesses.
[d]Total is more than sum because miscellaneous businesses have been omitted.

APPENDIX 14. **Exports realized directly by American cooperatives in 1976**

Products	Number of Exporting Cooperatives	Value of Exports (millions of $)	Share in the Total of United States Exports (%)
Animal products	22	34.2	1.4
Poultry products	16	16.5	6.3
Live animals and others	6	17.1	3.8
Grains	13	931.6	8.6
Wheat	4	356.2	9.2
Rice	6	85.0	13.5
Corn and sorghum	3	490.3	8.2
Fruits	35	372.2	38.4
Fresh citrus fruits	6	186.9	69.9
Other fresh fruits	9	10.3	6.2
Processed fruits	17	95.5	28.4
Fruits to be processed	3	79.5	40.1
Animal feed	76	10.1	2.3
Vegetables	12	18.4	2.7
Oilseeds	11	427.2	8.4
Cotton	4	231.7	22.2

Source: Donald Hirsh, "Cooperations Directly Export 2 billions $ in Farm Products," *Farmer Cooperations,* May 1978, cited in *Agriculture Cooperation,* see sources of Appendix 13.

APPENDIX 15. Evolution of the average productivity in American agriculture before and after the 1973 crisis

1967 = 100	Index Numbers			Average Annual Rate (%)	
	1959	1972	1979	72/59	79/72
1. Farm output	88	110	129	+1.7	+2.3
2. Total input	97	100	108	+0.2	+0.1
3. Labor	144	82	66	−4.2	−3.1
4. Farm real estate	99	98	96	+0.1	+0.3
5. Agricultural chemicals	75	101	182	+2.4	+8.8
6. Mechanical power and machinery	94	98	129	+6.3	+4.0
7. Global productivity (1/2)	91	110	119	+1.5	+1.2
8. Labor productivity (1/3)	61	134	196	+6.2	+5.5
9. Land productivity (1/4)	89	112	134	+1.8	+2.6
10. Productivity of fertilizers and pesticides (1/5)	117	109	71	−0.5	−6.0
11. Machine productivity (1/6)	93	112	100	+1.4	−1.6

Sources: USDA, *Agricultural Statistics, 1977 and 1980,* Tables 629 to 631, "Agricultural Productivity," "Total Farm Input," "Farm Production and Output."

Note: Productivity is the output-input ratio. A negative average annual rate of growth for productivity (chemicals and machinery, for example) means U.S. agriculture has gone into the productivity-decreasing phase for these inputs; it does not mean that the use of these inputs will not grow in the future.

APPENDIX 16. The sixteen largest user markets of soy products in order of decreasing importance of their imports in 1979–1980 in soy meal equivalent (SME)[a] (millions of tons)

	1977–78[b]			1978–79[b]			1979–80[b] (projection)		
	Beans	Meal	SME	Beans	Meal	SME	Beans	Meal	SME
1. West Germany	3.3	1.0	3.5	3.6	1.5	4.4	3.8	1.6	4.5
2. Japan	3.6	0.3	3.2	4.0	0.4	3.5	4.2	0.4	3.7
3. France	0.6	1.6	2.0	0.7	2.2	2.7	0.8	2.2	2.8
4. Netherlands	1.6	0.9	2.2	2.3	0.9	2.7	2.4	0.9	2.8
5. Spain	1.8	0.5	1.9	2.1	0.4	2.1	2.4	0.4	2.3
6. Italy	1.2	0.8	1.7	1.3	0.9	1.9	1.4	0.9	2.0
7. Great Britain	1.1	0.3	1.1	1.4	0.3	1.4	1.4	0.3	1.4
8. USSR (calendar years)	1.5	0.0	1.2	0.9	0.0	0.7	1.6	0.0	1.3
9. Belgium	0.7	0.4	1.0	1.0	0.5	1.3	1.0	0.6	1.3
10. Denmark	0.4	0.6	0.9	0.5	0.7	1.0	0.5	0.7	1.1
11. Taiwan	0.7	0.0	0.6	0.9	0.0	0.8	1.3	0.0	1.1
12. Poland	0.0	0.6	0.6	0.1	0.7	0.8	0.1	0.8	0.9
13. Mexico	0.6	0.2	0.6	0.6	0.1	0.6	0.9	0.0	0.8
14. South Korea	0.2	–	0.1	0.2	0.0	0.2	0.4	0.1	0.4
15. Yugoslavia	0.1	0.2	0.3	0.3	0.2	0.4	0.3	0.2	0.4
16. Continental China	0.3	0.0	0.2	0.2	0.0	0.1	0.4	0.0	0.3
World Total	18.4	9.0	23.6	22.5	12.1	30.0	24.7	12.0	31.7

Source: USDA News—Weekly Round Up of World Production and Trade, June 14, 1979.
[a]SME = tonnage of imported beans × 0.795 + tonnage of imported soy meal.
[b]Years beginning October 1.

APPENDIX 17. PL 480: Sales with special credits, 1979–1980

1. Countries with Per Capita Incomes Less Than $580	Total (millions of dollars)	Wheat	Rice	Corn-sorghum	Oil (1,000 t)	Food Total (1,000 t)	Cotton (thousands of bales)	Tobacco (1,000 t)
			(1,000 t)					
Bangladesh	62.1	400	—	—	15	415	—	—
Bolivia	12.0	93	—	—	—	93	—	—
Egypt	214.0	1,500	—	—	—	1,500	—	—
Guinea	6.0	11	10	—	2	23	—	—
Guyana	2.3	3	—	—	2	5	—	0.1
Haiti	16.7	30	12	—	15	57	—	—
Honduras	2.0	15	—	—	—	15	—	—
Indonesia	104.3	272	240	—	—	512	—	—
Madagascar	1.4	—	5	—	—	5	—	—
Morocco	0.8	84	—	—	—	84	—	—
Pakistan	40.0	260	—	—	10	270	—	0.4
Philippines	10.0	25	—	—	—	25	14	—
Sierra Leone	1.2	5	—	5	—	10	—	—
Somalia	10.7	24	10	24	3	61	—	—
Sri Lanka	17.0	133	—	—	—	133	—	—
Sudan	20.0	153	—	—	—	153	—	—
Zaire	17.0	22	30	—	—	52	10	0.5
Zambia	10.0	33	5	—	7	45	—	—
Ghana	10.0	48	10	10	—	68	—	—
Mozambique	5.0	17	10	—	—	27	—	—
Total 1	572.5	3,128	332	39	54	3,553	24	1.0
Total 1/Total	(78%)	(87%)	(76%)	(9%)	(84%)	(78%)	(38%)	(53%)

2. Countries with Per Capita Incomes Higher Than $580	Total (millions of dollars)	Wheat	Rice	Corn-sorghum	Oil (1,000 t)	Food Total (1,000 t)	Cotton (1,000 t)	Tobacco (1,000 t)
		(millions of tons)						
Dominican Republic	15.0	28	–	40	10	78	–	–
Israel	5.4	40	–	–	–	40	–	–
Jamaica	10.0	42	–	45	–	87	–	–
Jordan	5.4	40	–	–	–	40	–	–
South Korea	40.0	142	–	80	–	222	40	–
Mauritius Island	2.8	–	10	–	–	10	–	–
Peru	20.0	–	71	–	–	71	–	–
Portugal	40.0	70	10	202	–	282	–	0.4
Syria	12.1	45	15	–	–	60	–	0.5
Tunisia	12.0	60	–	30	–	90	–	–
Total 2	162.7	467	106	397	10	980	40	0.9
Total 2/Total	(22%)	(13%)	(24%)	(91%)	(16%)	(22%)	(62%)	(47%)
Not attributed	49.8	255	12	–	–	267	–	–
Overall Total	785.0	3,850	450	436	64	4,800	64	1.9

Source: USDA News, January 2, 1979.

APPENDIX 18. Medium-term export credits program: Beneficiary countries (millions of dollars)

Country	Amount Granted from 10/1/77 to 8/31/78	Amount Accumulated from 3/30/56 to 8/31/78	Major Farm Products Financed During the Fiscal Year 1978
Poland	441.3	1,023.7	corn, wheat, soybeans
Portugal	144.9	314.5	corn, wheat, soybeans
South Korea	128.4	1,230.0	cotton, wheat
Greece	76.1	302.0	corn
Pakistan	76.2	227.0	wheat, soy oil
Peru	59.5	319.2	wheat, corn, soy oil
Philippines	53.5	342.5	wheat, cotton, tobacco
Chile	23.3	83.5	wheat, corn
Romania	19.4	177.6	soybeans
Thailand	14.5	86.2	tobacco, cotton
United Kingdom	12.0	236.5	tobacco
Bolivia	12.0	31.7	tobacco, corn, rice
Yugoslavia	9.4	179.6	soybeans
Cyprus	9.2	21.4	sorghum, wheat
South Africa	8.3	95.9	rice, soy protein
Australia	5.1	32.0	tobacco
New Zealand	3.5	20.8	tobacco
Ireland	3.4	39.6	tobacco
Belgium	1.3	26.0	tobacco
Norway	0.8	29.7	tobacco
Costa Rica	0.1	0.1	breeding pigs and livestock
Spain	0.05	33.2	dairy cows
Total	1,342.4	4,852.7	

Source: The French agricultural attaché in Washington.

APPENDIX 19. Soviet grain balance sheet, 1972–1985

From July 1 to June 30	Production	Imports	Exports	Available	Usage						Stock Variation
					Total	Seeds	Industry	Food	Losses	Animals	
1972–73	168	22.8	1.8	189	187	26	3	45	15	98	+2
1973–74	223	11.3	6.1	228	214	27	3	45	33	105	+14
1974–75	196	5.7	5.3	196	206	28	3	45	23	107	−10
1975–76	140	26.1	0.7	166	180	28	3	45	14	89	−14
1976–77	224	11.0	3.3	232	221	29	3	45	31	112	+11
1977–78	196	18.9	2.3	213	228	28	4	45	29	122	−16
1978–79	237	15.6	2.8	250	231	28	4	46	28	125	+19
1979–80	179	31.0[a]	0.8	209	222	28	4	46	22	123	−13
1980–81	189	34.8	0.5	223	225	27	4	47	28	119	−2
1981–82	160	46.0	0.5	206	210	27	4	47	16	116	−4
1982–83	180	32.5	0.5	212	213	27	4	47	18	117	−1
1983–84	191	32.9	0.5	222	222	27	4	47	21	123	0
1984–85	170	57.0	1.0	226	222	27	4	48	19	124	+4
1985–86	191	32.0	1.0	222	223	27	4	48	20	124	−1

Source: USSR Task Force, USDA, Foreign Agricultural Service, Foreign Agricultural Circular.
[a]Year of the embargo.

After eleven months of debate and a ten-day House-Senate Conference, the Congress of the United States approved on December 18, 1985, and President Reagan signed on December 23, 1985, a five-year farm bill called the Food Security Act of 1985. It covers the period from 1986 to 1990. Although this compromise did not achieve the savings Mr. Reagan wished, it contained a number of strategic changes. For the first time since the historical decisions taken under the Kennedy-Johnson administration in 1960–64, support prices for wheat, corn, soybeans, rice, and cotton were lowered (sugar and tobacco are exempted).

As stated by Secretary of Agriculture John Block, the objectives of the administration were

- to lower the price-support loan rate in order to fix its level at 75 percent of the average world price;
- to discontinue the target-price and deficiency payments system in 1990 and to reduce it thereafter 20 percent a year from 1985 to 1990;
- to discontinue the farmers-owned reserve system;
- to increase the export-oriented expenses and decrease the income and price-support-oriented expenses in order not to spend more than $12 billion a year for these purposes;
- to scale down from $50,000 to $20,000 the limit of subsidies that a single farm could get from the government.

Once again, the results pleased few in Congress or the administration but, more importantly, they angered fewer. Two pragmatic mediators emerged as key writers of the bill: Congressman Thomas Foley, Democrat from Washington state and majority whip, and Senator Robert Dole, Republican from Kansas and majority leader, both of whom represent wheat growing regions.[1]

The Foley-Dole agreement cuts the loan rates while freezing the target prices. This will widen the gap between market prices (which follow the decreasing loan rate) and target prices, which means the government will have to pay more in income subsidies. Instead of $12 billion, the cost of income and price support will be $17 billion a year for the first three years, then will decrease only after 1988 when the target price will begin to decline.

In fact, the first decision of the USDA following adoption of the new farm bill was to put the wheat loan rate at $88 per ton ($2.40 per bushel), which represents a 27 percent decrease; for corn the decrease is 25 percent, from $100 per ton to $75 (from $2.55 per bushel to $1.92). The wheat acreage reduction is to be at least 25 percent, of which only 2.5 percent will be subsidized by payments in kind. If the acreage reduction decided by the farmer is 45 percent, another 10 percent will be subsidized.

These new loan rates, as decided by the USDA on January 13, 1986, are much lower than the world prices measured on the same day by the futures prices on the Chicago Board of Trade: $120 per ton for March wheat, $101 for July wheat; $98 for March and July corn futures prices.

The provisions of the Food Security Act (FSA) require the government to set the loan rate between 75 and 85 percent of the average domestic market price for the crops

1. David Rapp, "Farm Bill Offers Limited Win for All Sides" and "Major Provisions of Farm Bill Conference Report," *Congressional Quarterly,* Washington, D.C., December 21, 1985.

of three of the past five years (disregarding the highest and lowest years), as hoped by Secretary Block. But Congress added in the law that the basic loan rate could not be reduced by more than 5 percent from the previous year, except if the average market price in the previous marketing year is less than 110 percent of the loan rate for that year, or if it is necessary to provide competitive prices on the world market: in that case, the secretary of agriculture could reduce it by up to 20 percent (Findley Amendment), with a minimum of 10 percent in 1986. Deciding to put the loan rate at $88 per ton for wheat and $75 for corn means the 20 percent maximum limit has been set and as early as 1986 the 75–85 percent of world prices bracket will have been reached. This will give grain exports of U.S. origin an immediate impetus, making them competitive with Canadian, Australian, and Argentinian grains.

As far as acreage reductions are concerned, the January 13, 1986, decision put them also at the maximum allowed by the new law. The FSA sets acreage reduction requirements at a minimum of 15 percent of farms' established wheat bases in 1986 (and 20 percent in 1987 through 1990) if surplus stock exceeds 27 million tons (1 billion bushels). The secretary has the authority to increase the reduction without additional payments to a maximum of 25 percent in 1986, 27.5 percent in 1987, and 30 percent in 1988–1990, but if maximum limits are set in 1986, farmers would receive the equivalent of 2.5 percent of payment in kind by government-owned commodities.

If final surplus stocks of feed grains exceed 50 million tons (2 billion bushels), the minimum acreage reduction would be 12.5 percent in 1986 through 1990 and the maximum without additional payments would be 20 percent.

This new mandatory set-aside policy is intended to avoid the creation of surplus stock. Its success will depend on the financial benefits which the farmers may expect from signing in with the USDA. On one side, the existing $50,000 limit on deficiency payments to individual producers has been maintained (and not reduced to $20,000 as the Reagan administration had hoped). But some payments will now be exempted from the ceiling: 1) the loan-rate payments when they are reduced by more than 5 percent a year; 2) the gains realized when crop loans are repaid at a rate equal to the prevailing world market price when it is lower (but not less than 70 percent of the loan rate) than the effective loan rate (this is a new concept called "marketing loan" introduced in the FSA); 3) the deficiency payments due if a producer devotes more acres than required to approved conservation uses or nonprogram crops.

On the other side, the amount of money which could be paid to farmers under FSA is important for the first two years 1986 and 1987, when target prices are frozen at their existing level ($161 per ton for wheat, $119 for corn), and even later, for they will be only slightly diminished—by 2, 3, and 5 percent—in the three following years. The deficiency payments are in proportion with the difference between the target price and the loan rate which means they will grow 1) for wheat, from $40 per ton in 1985 ($161 minus $121) to $73 in 1986 ($161 minus $88) and possibly to $75 in 1990 ($145 minus $70); 2) for corn, from $18 per ton in 1985 ($119 minus $101) to $24 in 1986 ($119 minus $95), and possibly to $48 in 1990 ($108 minus $60). With a current limit of $50,000 per farm and a current yield of 3 tons per hectare of wheat and 6 tons per hectare of corn, this means wheat farms of more than 228 hectares (570 acres) and corn farms of more than 347 hectares (868 acres) will not be induced to participate in the program in 1986.

Another consequence of FSA could be to favor lower world prices. The decreased loan rate together with the new marketing loan system could permit world prices to be set at a level of 70 percent of reduced loan rates without being higher than U.S. prices. This could lead to world prices of $50 per ton for wheat and $40 per ton for corn in 1990, that is, less than half of the January 1986 world prices considered already as low.[2]

To prepare U.S. agriculture for that eventuality, the FSA established for ten to fifteen years a long-term Conservation Reservation Program for at least 16 million hectares (40 million acres) and up to 18 million hectares (45 million acres) of erosion-prone land; this target will be reached in the next five years with USDA aid covering up to 50 percent of the cost of installing approved cover-crops. The FSA also discourages cultivation of highly erodible land (sod buster and swamp buster programs) by excluding those lands from price and income support programs and from insurance protection.

Also to prepare U.S. agriculture for a long struggle in the world markets, a strong export financing program is implemented by FSA as amended on March 20, 1986: 1) the short-term (six to three years) export credit guarantees (GSM 102) will be made available at a minimum level of $5 billion a year; 2) the intermediate-term (three to ten years) export credit guarantees (GSM 301) will be not less than $500 million a year and not more than $1 billion a year and may be used for food exports to the most indebted developing countries (Baker Plan); 3) an Export Enhancement Program will continue the BICEP program (Bonus Incentive Commodity Program or GSM 500): $333 million of CCC-owned commodities will be provided each year in order to give an export bonus to subsidized markets; 4) an Export Assistance Program will continue the mixed credit program (GSM 5): $110 million a year of CCC funds or CCC commodities will be used to promote U.S. commodities adversely affected by the price or credit subsidies or unfair marketing arrangements used by other countries; 5) the Food for Peace Program (PL 480) is extended with its existing ceiling for grant programs at $1 billion a year and 1.8 million metric tons, increased to 1.9 in 1989 and 1990 of which 75 percent at least will be donated in the form of transformed or enriched or packed products; and 6) there will be also a new Food for Progress Program to promote a private free enterprise policy in recipient countries (75,000 to 500,000 tons of grain a year).

What is new in this export policy is mandatory spending of export credits and subsidies. When U.S. wheat exports are down in 1985/86 to 26 million tons from 39 million in 1984/85 and wheat closing stocks are up on June 30, 1986, to 49 Mt from 39 Mt on June 30, 1985, it is understandable that the Dole-Foley compromise relies on mandatory export spending. The picture of the feed grain situation is as bad: exports are down to 49 Mt from 56 and closing stocks up to 102 Mt from 50 Mt. Only soybean exports are increasing—to 19 Mt from 16 Mt, but closing stocks are increasing to 16 Mt on August 31, 1986, from 9 Mt on August 31, 1985. The U.S. market share of world exports for wheat was 48 percent in 1982 and 36 percent in 1985, for feed grains it was 61 percent in 1982, 55 percent in 1985, and 52 percent in 1986. For total food and agriculture exports the U.S. market share has decreased from the record 25 percent in 1974 to 19 percent in 1980 and to 15 percent in 1983; in value, the U.S. food and agriculture exports have been down from $38 billion in 1984 (balance plus 19) to $34 billion in 1985 (balance plus 14).

Farm income is also dropping. The farm crisis is mainly affecting the middle-size farms, which have sales between $40,000 and $100,000. Only 15 percent of the 2.3

2. Ralph Ichter and Olivier Dubuquois, "Le Farm Bill de 1985," Note of the French Agricultural Attaché. (Embassy of France, Washington, D.C., December 16 and 19, 1985.)

million U.S. farms are making a profit, and 15 percent are near bankruptcy. This is why the FSA is providing new credits for the Farmers' Home Administration reserved for family farms ($4 billion a year). The FSA is also linked to a bill to rescue the Farm Credit System: a new three-member board of directors, all appointed by the president, replaces the board elected by the thirteen representatives from the banking districts and will redistribute the $7 billion surplus reserve throughout the system and take over bad loans; if needed, the USDA budget will pump additional money into the system.

The 1985 farm bill would not be complete without a new dairy policy; in fact, there will be an 18-month program to remove 800,000 of the nation's 11 million dairy cows from milk production (7 percent of production). This program will be financed by a 4 percent assessment on milk prices; the assessment and price cuts combined will decrease the effective price support of $11.60 per hundred pounds by 4 percent in 1986, 3 percent in 1987, and 5 percent the next years until the government purchases do not exceed 5 billion pounds a year (16.5 billion pounds were purchased in 1985).

As it is written the Food Security Act could be countered by the Gramm-Rudman Balanced Budget Law, which is asking for a 4.3 percent reduction from 1985 to 1986 in USDA spending. In this event, nobody knows how the freeze of the target prices could be maintained, even with the provision giving the secretary discretion to issue export certificates to wheat and feed grain producers with a cash value and to redeem them from exporters, giving producers a possible alternative source of income if the secretary were to reduce deficiency payments by 13 cents a bushel for wheat and 6 cents a bushel for corn. The complexity of this measure, which could be understood as an export tax, seems to prevent its use.

This new farm bill, even if complex and sometimes obscure, will give a new boost to U.S. agricultural exports, which at the same time will benefit from the dropping rate of exchange of the U.S. dollar. But nothing is really clear and the budgetary cost could explode. This prospect might explain why, after the vote of the 1985 farm bill, Secretary Block chose to resign and has been replaced by Richard Lyng, his former undersecretary at the USDA.

Notes

CHAPTER 1: GLOBAL BALANCE AND PARTIAL IMBALANCES THROUGH THE YEAR 2000

1. This is the problem of "simultaneity," well known to economists. With respect to agricultural markets, this phenomenon presents a number of special characteristics associated with the lag between planting and harvesting and with the annual rhythm of most crops.

2. FAO, *Assessment of the World Food Situation Present and Future,* prepared and published for the United Nations World Food Conference, Rome, November 5–16, 1974; and University of California, *A Hungry World: The Challenge to Agriculture* (1970).

3. OECD, *Etude de tendances de l'offre et de la demande des principaux produits agricoles* (Paris: OECD, 1976); and USDA, Economic Research Service, *The World Food Situation and Prospects to 1985* (Washington, D.C.: USDA, 1973), completed by the Economic, Statistics, and Cooperative Service, *Alternative Futures for World Food in 1985, World GOL Model Structure and Equations,* Foreign Agricultural Economic Reports 146 and 151 (Washington, D.C.: USDA, 1978).

4. L. Blakeslee, E. Heady, and C. Framington, *World Food Production and Trade* (Ames: Center for Agricultural and Rural Development, University of Iowa, 1973); and International Food Policy Research Institute, *Meeting Food Needs in the Developing World* (Washington, D.C.: IFPRI, 1976), completed by *Food Needs of Developing Countries—Projections of Production and Consumption to 1990,* Research Report 3 (Washington, D.C.: IFPRI, 1977).

5. A Study by GEPI (Groupe d'Etude des Politiques Internationales) had already shown the weaknesses of the FAO's 1974 analysis: *Problèmes actuels et perspectives de la repartition alimentaire mondiale* (Paris: Commissariat General du Plan, 1974), and FAO has progressively modified its outlook in *Agricultural Commodity Projections, 1975–1985* (Rome: FAO, 1979), then in *Agriculture = toward 2000* (Rome: FAO, 1981), and finally in unpublished reports for the tenth anniversary of the World Food Conference (Bellagio, February 1984 and Addis Ababa, April 1984).

6. Winrock International, *World Agriculture Review and Prospects into the 1990's* (Morrilton, Ark.: W. I. Livestock Research and Training Center, 1984). This six hundred-page study was prepared under the direction of Howard Yorth, former director of economic affairs for the USDA under the Carter administration. Resources for the

Future, in cooperation with Economic Perspectives, Inc., *Food and Fiber Projections to 2000* (Washington, D.C.: Joint Council on Food and Agricultural Sciences, Reference Document, 1984). This 460-page study has been developed by Fred H. Sanderson, senior fellow at Resources for the Future, who presented a communication to the International Association of Agricultural Economists, "Food Demand, Production and Trade in the Year 2000" (Kiel, 1984).

7. International Wheat Council, *Long-Term Grain Outlook*, Secretariat paper 14 (London: International Wheat Council, 1983).

8. The National Center for Food and Agricultural Policy conducted preparatory studies for the revision of the GOL model.

9. *World Food Study for the Year 2000* was produced at the USDA by a team directed by Clark Edwards. Also see Vernon Roningen and Karen Liu, *The World GOL Model: Background and Standard Component* (Washington, D.C.: USDA, Economic Research Service, 1983).

10. U.N. Fund for Population Activities, *Demographic Indicators of Countries, Estimates and Projections as Assessed in 1980* (New York: UNO, 1982); and Rafael Salas, "Second World Conference on Population," (Mexico City, 1984). See also World Bank, *World Development Report 1984* (Washington, D.C.: World Bank, 1984), which is mainly devoted to population growth problems.

11. Robert S. MacNamara, address to the Council of Governors in Belgrade, Yugoslavia, October 2, 1979 (Paris: World Bank, European Bureau, 1979).

12. See Appendix 2 for a more detailed analysis of this complex idea of yields in units per hectare for a given farm.

13. In India, it was estimated that more than 30 percent of annual stocks were destroyed by bad storage conditions in the 1970s.

14. FAO, *Trade Yearbook* (Rome: FAO, various years).

15. The U.S. Congress directed the USDA to assess the needs of every developing country regularly, taking into account its projected production. The *FANA* or "Food Availability and Needs Assessment" is prepared annually by the Economic Research Service for USAID.

16. In other words, when real income increases.

17. Gale D. Johnson, *World Agriculture in Disarray* (London: Macmillan/Saint Martin Press, 1973). The FAO study has not been published but is used in the OECD study, *Etude de tendances de l'offre et de la demande*. C. Riboud, "Welfare Effects of Farm Commodity Programs in the U.S.: 1948–1973" (Ph.D. dissertation, Massachusetts Institute of Technology, 1978).

18. MacNamara, address, October 2, 1979.

19. Bernard Auberger, *Rapport protéine* (Paris: BIMA [Ministry of Agriculture], 1977).

20. MacNamara, address, October 2, 1979.

21. Z. Griliches, "The Sources of Measured Productivity Growth: U.S. Agriculture: 1940–1960," *Journal of Political Economy* 71 (August 1963): 331–46.

22. With respect to labor, we will note that the USDA series indicates a progression of nearly 300 percent of input per work unit between 1948 and 1973.

23. Y. Mundlak, "On the Pooling of Time Series and Cross Section Data," 1976, mimeo; L. Lau and P. Yotopoulos, "A Test for Relative Economic Efficiency: Some Further Results," *American Economic Review* (March 1973): 214–23. Unfortunately, this study, for the time being, uses econometric techniques which raise some doubts as to the validity of the authors' conclusions.

24. The rate of return for large World Bank projects continues to be a subject of active debate, particularly with respect to the agricultural sector.

CHAPTER 2: THE AGRICULTURAL DILEMMA FOR FRANCE AND EUROPE

1. Trade in food and agricultural products outside of the EEC in 1982 amounted to imports worth $72 billion and exports worth $64 billion (*Statistiques de base de la CEE* [Luxembourg: Commission of the European Community, Eurostat, 1983]).

2. *Annuaire de statistique agricole de la C.E.E. 1974–1977*, Table C1, "Nombre et superficie des exploitations agricoles par classes de grandeur" (Luxembourg: Commission of the European Community, Eurostat, 1980).

3. Fred H. Sanderson, "Food Demand, Production and Trade in the Year 2000" (Kiel: International Association of Agricultural Economists, 1984); Winrock International, *World Agricultural Review and Prospects into the 1990's* (Morrilton, Ark.: W. I. Livestock Research and Training Center, 1984).

4. Alan Holz, "World Oilseeds and Products Outlook, 1980" (Washington, D.C.: Agricultural Outlook Conference, November 7, 1979), Table 32.

5. Gale Johnson, professor at the University of Chicago, estimates at $1 billion the excess consumption of soy meal in the EEC resulting from this relative price distortion. This would represent about 5 Mt of soy meal out of the 15 Mt imported in 1979 by the European Community. Of course, the reduction in imports of soy products would result in greater use of grain in livestock rations (see Gale Johnson et al., *World Agriculture and Trade Policies: Impact on U.S. Agriculture* [Washington, D.C.: American Enterprise Institute for Public Policy Research, 1979]).

6. See Jean Abonnenc, president of the Syndicat National des Industriels de l'Alimentation Animale, report presented to CENAG on June 19, 1979: "Céréales, armes stratégiques."

CHAPTER 3: THE DOMINANT ROLE OF THE UNITED STATES IN THE WORLD AGRICULTURAL ECONOMY

1. Last date for which FAO figures were available.

2. C. Yeh, L. Tweeten, and L. Quance, "U.S. Agricultural Production Capacity," *American Journal of Agricultural Economics* 59 (February 1977): 37–48.

3. The total of world exports in wheat, rice, oilseeds, fats (excluding butter), and oil meal as a percentage of the value of world agricultural exports, excluding fishing and forestry products (FAO *Trade Yearbooks*).

4. L. Tweeten and F. Tyner, "Excess Capacity in U.S. Agriculture," *Agricultural Economic Research* 16 (January 1964): 23–31; L. Quance and L. Tweeten, "Excess Capacity and Adjustment Potential in U.S. Agriculture," *Agricultural Economic Research* 24 (July 1972): 57–66.

5. H. Spielman and E. E. Weeks, "Inventory and Critique of Estimates of U.S. Agricultural Capacity," *American Journal of Agricultural Economics* 57 (December 5, 1975): 923–28.

6. These problems do not prevent an estimation of the supply and demand functions in the context of disequilibrium. See on this subject C. Riboud, "Intervention du gouvernement et modèles économétriques de l'agriculture," Working document of the Economics Seminar of M. Edmond Malinvaud, December 1980.

7. See on this subject, L. Johansen, *Production Functions* and K. Sato, *Aggregate Production Functions* (North Holland Publishing Company).

8. Yeh, Tweeten, and Quance, "U.S. Agricultural Production Capacity."

9. See Table 1.1 in Chapter 1; the "basic" scenario is similar to the status quo scenario of the GOL model.

10. M. L. Cotner, M. O. Skold, and O. Krause, *Farmland: Will There Be Enough?* ERS, 584 (Washington, D.C.: USDA, 1975).

11. FAO, *Production Yearbook, 1979*, Table 97.

12. P. Crosson, "U.S. Agricultural Production Capacity: Comment," *American Journal of Agricultural Economics* 60 (February 1978): 144–47.

13. Statistics from the World Wheat Council.

14. See the excellent work by Dan Morgan, *Merchants of Grain: The Power and Profits of the Five Great Companies at the Center of the World's Food Supply* (New York: Viking Press, 1979), and in particular the maps on the frontispiece page. The French translation was published by Fayard under the title *Les Géants du grain*.

15. On the difficulties of Soviet agriculture, see T. Gustavson, *Reform and Power in Soviet Politics: Lessons of Brezhnev's Agricultural and Environmental Policies* (Cambridge: Cambridge University Press, 1981).

16. During the second week of July 1980, India imported 25,000 metric tons of American wheat. These were the first shipments since 1976. By contrast, India exported 0.5 million tons of wheat in 1977, 0.4 million tons of rice in 1978, and 0.3 million tons of rice in 1979. In 1980, India finally was a net exporter of 0.3 million tons of grain. In 1981, 1982, and 1983 the country was a net importer of 9.3 million tons of grain and in 1984 and 1985 a net exporter of 3.5 million tons of grain.

17. The Williams report, named for Albert M. Williams, chief of finance for IBM, the chairman of the study group appointed by President Nixon, on which Ken Ogren, former U.S. agricultural attaché in Paris and at the OECD, served as head of the agricultural report committee, was published in two volumes: Commission in International Trade and Investment Policy, *United States International Economic Policy in an Interdependent World: Report to the President* (Washington, D.C.: U.S. Government Printing Office, 1971). The Flanagan report, named after the director of the Council of International Economic Policy, created in 1972, was written in 1973 and published under the title *Agricultural Trade and the Proposed Round of Multilateral Negotiations* by the Committee on Agriculture and Forestry, U.S. Senate, 93d Cong., 1st sess. Some of the ideas publicized by these two reports had already been proposed in the Berg report: National Advisory Commission on Food and Fiber, Sherwood O. Berg, Chairman, *Food and Fiber for the Future: Report to the President* (Washington, D.C.: U.S. Government Printing Office, 1967).

18. Equal to half the value of oil purchases assessed in early 1979.

19. See Jacques Poly, director-general of the INRA, "Le rôle de la recherche agricole face aux nouveaux problèmes de l'agriculture" (FAO, Agricultural Commission for Europe, 1980).

20. Turner Oyloe, former assistant director of FAS and later agricultural counselor in Paris, projects an increase of 66 percent in grain and oilseed exports between 1980 and 1990 ("U.S. Agricultural Exports in the 1980's," Outlook Conference, Washington, D.C., November 6, 1979).

21. Planning and Analyzing Staff, Office of the Administrator of the Foreign Agricultural Service, *International Trade and Financial Flows: Implications for Credit and Aid*, FAS Staff Report No. 2 (Washington, D.C.: USDA, 1984).

22. USDA, *Agricultural Food Policy Review*, Economic Research Service, AFPR-1 (Washington, D.C.: USDA, 1977).

23. This voluntary policy was accompanied by conceptual revisions related to the operation of international markets; see C. Riboud, "Le modèle GOL," *Revue de politique étrangère*, no. 2 (Spring 1980).

CHAPTER 4: THE OBJECTIVES AND INSTRUMENTS OF AMERICAN AGRICULTURAL POLICY

1. R. Talbot and D. Hadwiger, *The Policy Process in American Agriculture* (San Francisco: Chandler, 1968).

2. See the remarkable article by M. Weitzman, "Prices vs. Quantity," *Review of Economic Studies*. Many articles followed on the same or connected subjects (tariffs versus quotas in international trade).

3. W. Cochrane and M. Ryan, *American Farm Policy, 1948–1973* (Minneapolis: University of Minnesota Press, 1976).

4. See C. Riboud, "Welfare Effects of Farm Commodity Programs in the U.S.: 1948–1973" (Ph.D. dissertation, Massachusetts Institute of Technology, 1978).

5. See, for example, the work cited in note 11, which gathered opinions prior to the preparation of the Agricultural and Consumer Protection Act of 1977 by the Senate Agriculture Committee.

6. Each representative is entitled to an office staffed by three professionals and each senator to an office with nine professionals. Each Senate and House committee employs about thirty people.

7. Spitze, "Future Directions for Agricultural Policy," USDA, ERS *Bulletin* (1977).

8. See Christophe Riboud, "Analyse économique des instruments d'une politique agricole" (Laboratoire d'économie politique de l'école normale supérieure et Centre d'analyse et de prévision du ministre des affaires étrangères, Paris, 1978).

9. Luther Tweeten, professor of agricultural economy at the University of Oklahoma, wrote a very interesting article, "Agricultural Policy: A Review of Legislation, Programs and Policy," which appeared in *Food and Agricultural Policy*, texts from the conference, organized by the American Enterprise Institute for Public Policy Research, on March 10, 1977.

10. Since the EEC grain settlement of 1962, ONIC has lost its monopoly on the purchase of French grain.

11. Wayne Rasmussen, Cladys Baker, and James Ward, "A Short History of Agricultural Adjustment, 1933–1975," article prepared by the Economic Research Service for the Senate Agriculture Committee and published in September 1976 in the Senate document, *Farm and Food Policy, 1977*.

12. a: Sales for a normal harvest $= 0.70 \, p_p \times \bar{q}$
b: Sales for a poor harvest $= 0.90 \, p_p \times 0.75 \, \bar{q} = 0.675 \, p_p \times \bar{q}$
c: Sales for a large harvest $= 0.60 \, p_p \times 1.30 \, \bar{q} = 0.780 \, p_p \times \bar{q}$
The normal sales are reduced in *b* and increased in *c* to account for the variable expenses depending on size of the harvest since the point is to stabilize parity income and not sales.

13. Tweeten, "Agricultural Policy."

14. See Alain Revel and Emmanuel Drion, "La nouvelle loi-cadre agricole des

Etats-Unis," in the *Information Bulletin* by the French agricultural attaché for the United States and Canada, January 1971.

15. Ibid.

16. There are two types of disaster payments: (1) payments for the inability to plant: 33.3 percent of the target price on 75 percent of the land area concerned (in 1978: $3 million for wheat, $7 million for corn); (2) payments for below normal yields: 50 percent of the target price for production deficits below 60 percent of the normal yield (in 1978: $86 million for wheat and $81 million for corn). But this program, designed in 1973, has been criticized by the Reagan administration, which wants to replace it by a national agricultural insurance system mixing the Federal Crop Insurance Act and the Low Yield Disaster Payment Program. That new program would apply to all the crops of a farm and would be operated by private insurance companies.

17. Economic Research Service, *Economic Indicators of the Farm Sector* (Washington, D.C.: USDA, 1983).

18. Tweeten, "Agricultural Policy."

19. Ronald Brownstein, "In Era of Record Deficits, Farm Price Supports Seem Likely Target for Cuts," *National Journal*, February 11, 1984.

20. Richard Henry, "Paiement en nature: Les ajustements de la politique agricole américaine," *SEDEIS* (March 1984), and Doug Jackson, "The U.S. P.I.K. Program: Perspectives for 1983" (London, 1983).

21. James G. Vertrees and Andrew A. Morton, *Crop Price-Support Programs: Policy Options for Contemporary Agriculture* (Washington, D.C.: U.S. Congress, Congressional Budget Office, 1984).

22. Carol Brookins and Catherine Hay, "Agricultural Policy," *World Perspectives*, Washington, D.C., December 1984.

23. USDA, *Agricultural Statistics 1982*, Table 548. The new definition of agricultural operations adopted in 1978 (see USDA, *Agricultural Statistics, 1979*, Table 600) does not permit a comparison of the figures given for the years prior to 1975. The new definition results in a decrease of 10 percent of the total number and an equal increase in the average size. For a more dramatic approach to this income distribution problem, concentrating on net farm income and neglecting off-farm income, see James Wessel with Mort Hantman, *Trading the Future: The Concentration of Economic Power in Our Food System* (San Francisco: Institute for Food Development Policy, 1983).

24. These calculations were made by Lary Wipf and Bela Balassa, World Bank economists, and were used in *U.S. Agriculture in a World Context: Policies and Approaches for the Next Decade*, a series of articles collected by Gale Johnson, of the University of Chicago, and John Schnittker, former assistant secretary in the Kennedy administration (New York: Praeger and the Atlantic Council, 1974).

25. The report by the International Trade Subcommittee of the Senate Finance Committee, published in July 1979, for congressional debate on the ratification of the Tokyo Round GATT agreements, recognized that the EEC dominated the world cheese market with 37 percent of world exports, excluding intra-Community trade. New Zealand and Australia, the second and third largest, account for only 14 and 9 percent respectively.

26. For more information on the two prices of milk in the United States, see USDA, *Agricultural Statistics, 1982*, Table 502, "Federal Milk Orders Markets Class I Price," and Table 504, "Prices Paid by Plants for Manufacturing Grade Milk for Evaporated Milk, Butter and Cheese."

CHAPTER 5: THE AGRICULTURE INDUSTRY

1. Commission du Bilan, *La France en mai 1981, les activités productrices* (Paris: La Documentation Française, 1981), p. 116.

2. W. Sunderland (pesticides) and Thomas Gillett (fertilizers), *Industrial Outlook, 1978, for Agricultural Chemicals* (Washington, D.C.: U.S. Department of Commerce, 1978).

3. See USDA, *Agricultural Statistics, 1982,* Table 608. The 2.4 million members of the agricultural sector divided a total net farm income of $24 billion. Production expenses were $140 billion.

4. The following publications contain information on the various food industries: the *Industrial Outlook* series published annually by the U.S. Department of Commerce; Standard and Poor's Industry Surveys, *Beverages Basic Analysis*; and Alain Revel, "Le point sur l'agriculture américaine," *Génie rural* (Paris), February 1972.

5. Among the ten largest world companies for food and drink cited by the 1979 directory, *Eurofood World Directory of Food and Drinks Manufacturing Companies,* in descending order, were seven American companies: Unilever, Great Britain; Nestles, Switzerland; Beatrice Food, U.S.; General Food, U.S.; Esmark, U.S.; Kraft, U.S.; Greyhound, U.S.; Ralston-Purina, U.S.; Allied Breweries, Great Britain; Coca-Cola, U.S. Following numerous mergers (there were seven hundred in the United States in 1982) the ranking in 1984 was as follows: Nestle-Carnation, Switzerland, $15 billion in sales; Beatrice Foods-Esmark, U.S. ($13.5 billion); Procter and Gamble, U.S. ($12.5 billion); Dart and Kraft, U.S. ($10 billion).

6. Jean-Claude Trunel, "L'évolution des prix alimentaires aux Etats-Unis," note of the French agricultural attaché in Washington, May 1979.

7. Richard Gilmore, "American Producers' Interests and the Grain Trade," paper presented at the Conference on Agricultural Grain Marketing, Topeka, July 26, 1978. Gilmore was a staff member of the Senate subcommittee on Multinational Companies, who later published a book, *Poor Harvest: The Clash of Policies and Interests in the Grain Trade* (New York: Longman, 1982).

8. Dan Morgan, *Merchants of Grain: The Power and Profits of the Five Great Companies at the Center of the World's Food Supply* (New York: Viking Press, 1979; Penguin Books, 1980).

9. Alain Revel et al., "La difficile mise en oeuvre d'une politique d'exportation agricole, l'expérience des Pays-Bas et des Etats-Unis, les possibilités de la France," *Académie d'Agriculture de France,* February 1979.

10. Jean-Francois Mittaine, "Les contraintes du négoce international," *CENAG,* July–August 1979.

11. Yves Chavagne, "L'industrie de la faim," *Témoignage Chrétien,* January 15, 1979.

12. *1979 Handbook of Agricultural Charts* (Washington, D.C.: USDA, 1980), Chart 94, and *Graph-Agri 1979* (Paris: Ministère de l'Agriculture de la République Française, SCEES, 1980), p. 64.

13. USDA, *Agricultural Statistics,* 1979, Table 774, "Consumer Price Index."

14. Ibid., Table 644, "Marketing Spreads."

15. Louis Harris and Associates, "Consumerism at the Crossroads, 1976," and "Consumerism in the Eighties, 1982." These polls are referred to in *Economie et Consommation* (Paris), February 1984.

16. William Boehm and Paul Nelson, *Current Economic Research on Food Stamps Use* (Washington, D.C.: USDA, ERS, 1978), indicates that for $5 billion spent on the program, food expenditures increased by $2 billion, or 1 percent of the $219 billion spent on food by Americans in 1977.

17. Paul Nelson and John Perrin, *Economic Effects of the U.S. Food Stamp Program in 1972–1974*, USDA Economic Research Service, Agricultural Economic Report No. 331 (Washington, D.C.: USDA, 1976). The 40,000 jobs eliminated by these transfers in the service and luxury product sectors should be subtracted from the total number of jobs created (105,000).

18. K. L. Robinson, "Preserving Family Farms: Structural Policy Issues in the U.S." (Ithaca, N.Y.: Department of Agricultural Economics, Cornell University, November 1978).

19. USDA, *Agricultural Statistics, 1979*, Table 657, "Total Personal Income."

20. Henri de Farcy, *Un million d'agriculteurs à temps partiel?* (Paris: Le Centurion, 1979). See also Appendix 9.

21. Thomas Edmondson and Kenneth Krause, *State Regulation of Corporate Farming*, USDA, ESCS, AER, no. 19 (Washington, D.C.: USDA, 1978).

22. Ronald Mighell and William Hoofnagle, *Contract Production and Vertical Integration in Farming, 1960 and 1970*, ERSA, no. 474 (Washington, D.C.: USDA, 1972).

23. USDA, *Agricultural Statistics, 1982*, Table 644, "Marketing Spreads"; and Trunel, "L'évolution des prix alimentaires aux Etats-Unis." See also *Fact Book of Agriculture* (Washington, D.C.: USDA, 1984).

CHAPTER 6: THE AMERICAN MODEL

1. Henri de Farcy and René Groussard, "Où va l'agriculture française?" Communication to the French Academy of Moral and Political Sciences, May 19, 1980.

2. USDA, *Agricultural Statistics, 1977*, Tables 665 and 666; and *1982*, Tables 583 and 584.

3. Florence Jacquet and Albert Chominot, "Les tendances du coût de production de céréales aux E.U.: Le cas du blé," Communication on the American grain perspectives to the French Academy of Agriculture, Paris, May 18, 1983.

4. G. E. Schuh, economist from the University of Minnesota, working with the World Bank, reported by *Expo-Corn*, France—Maïs, Toulouse, April 1984.

5. Dan Morgan, *Merchants of Grain: The Power and Profits of the Five Great Companies at the Center of the World's Food Supply* (New York: Viking Press, 1979).

6. For information on corn blight, or Helminthosporium maydis race T, see R. Cassini, "Etat actuel des principales maladies du maïs," *Phytiatrie-Phytopharmacie*, no. 22 (Paris), 1973, and A. S. Ullstrup, "The Impacts of the Southern Corn Leaf Blight Epidemics of 1970–1971," *Journal Paper*, no. 4, 620 (Purdue University Agricultural Experiment Station, 1974).

7. Daniel Colon, "La place de l'huile de palme dans le marché mondial des corps gras" (Paris: *Oleagineux*, April 1979). The conversion rates, oil/seed, are as follows: soybeans, 18 percent oil per weight unit of seeds; cotton, 18 percent; sunflower, 35 percent; rapeseed, 38 percent; ground nuts, 46 percent; palm, 48 percent; copra, 64 percent.

8. A. Vidal, "La sélection du soja aux Etats-Unis," Plant Breeding Laboratory, unpublished mission report (Montpellier: INRA, 1976).

9. Section 812 added by the Agriculture and Consumer Protection Act of 1973 requires a declaration to the USDA by all exporters of weekly exports of seventeen listed products and any contracts signed during the week. If the quantities exceed 100,000 metric tons per day, or 200,000 metric tons per week, the declaration must be made daily. The information collected is published, with a delay of one week, in *U.S. Export Sales*. For more information see French agricultural attaché to Washington, "Le système d'information de l'U.S.D.A. sur les exportations," *Bulletin d'informations agricoles*, December 1979; and Office of the General Sales Manager, *A Report on U.S. Exports 1974–1978* (Washington, D.C.: USDA, 1978).

10. Michel Bernon, "Le tabac, aliment de demain," memo from the scientific adviser to the French Embassy in Washington, D.C., June 1979.

11. See Wayne Sharp, agricultural counselor to the U.S. Embassy in Paris, "La politique céréalière des U.S.A.: Arme stratégique ou outil de coopération internationale?" *CENAG* (Paris), June 20, 1979. Assuming world demand elasticity of −.5, Sharp studied the effects of an increase in world grain prices of 70 percent. World grain exports, 155 Mt in 1977–78, would be reduced by 35 percent, or 55 Mt, thus depriving the four main exporters of more than 40 percent of their exports. Income earned on these exports would hold steady because of the increase in price. The main victims of such a measure would be Japan, the Eastern bloc countries, and China. The EEC would suffer only mild budgetary problems and could then export its surplus grains (wheat and barley) without the use of export subsidies. The second consequence would affect American livestock breeders. Domestic grain consumption in the United States is 160 Mt, or about twice the amount exported. Using Sharp's assumption of domestic American demand elasticity of −.3, domestic consumption would decrease by 21 percent, or 34 Mt, and American food prices would rise by 15 to 20 percent. This would detrimentally affect the economic position of both U.S. farmers and consumers.

12. James Nix, "Outlook for Livestock and Meat," Outlook Conference, November 15, 1978.

13. F. A. Jaenke, "Agricultural Research, A High Priority," in *Farm and Food Policy 1977* (Washington, D.C.: U.S. Senate, 1976).

14. See Gustave Strain, "La politique de recherche agro-alimentaire américaine" (Scientific Mission to the French Embassy in Washington, April 1979). The National Agricultural Research Extension and Teaching Policy Act of 1977 created an Agricultural Science and Education Administration within the USDA.

15. See Jacques Poly, *Recherche agronomique, réalités et prospectives* (Paris: INRA, 1977).

16. Theodore W. Schultz, "The Value of the Ability to Deal with Disequilibria," *Journal of Economic Literature* 12 (September 1975): 841, 843.

17. In 1978 renamed the Economic Statistical and Cooperative Service (ESCS) and in 1980 again renamed ERS.

18. Since 1978, the Outlook Boards have been coordinated by a central board, the World Food and Agricultural Outlook and Situation Board.

19. Carol Brookins and Bernard Drury, "Task Group Reports on Crop Reporting Board Procedures," *World Perspectives*, August 1, 1984.

20. Consultative Group on International Agricultural Research, *An Integrative Report* (Washington, D.C.: World Bank, 1978).

21. Morgan, *Merchants of Grain*.

22. J. Hand, "The Demand for Agricultural Credit," in Polakoff et al., *Financial Institutions and Markets* (Boston: Houghton-Mifflin, 1970); A. G. Nelson and W. G.

Murray, *Agricultural Finance*, 5th ed. (Ames: Iowa State University Press, 1967); O. B. Quinn, "Sources and Uses of Funds in Agriculture," *American Journal of Agricultural Economics* 53 (February 1971).

23. Hand, "Demand for Agricultural Credit."

24. Quotes from the Federal Farm Loan Act of 1916.

25. Ludke, *The American Financial System: Markets and Institutions* (Boston: Allyn and Bacon, 1967).

26. G. L. Swackhamer and R. S. Doll, *Financing Modern Agriculture: Banking's Problems and Challenges* (Kansas City: Federal Reserve Bank of Kansas City, 1969).

27. *Food for Peace, 1977 Annual Report on Public Law 480*, sent to Congress by the secretary of agriculture, July 10, 1978.

28. Brady J. Deaton, associate professor, Department of Agricultural Economics, Virginia State University, and staff coordinator of USDA Special Task Force on P.L. 480, "Public Law 480: The Critical Choices" (American Agriculture Economics Association, 1980); Theodore W. Schultz, *The Economics of Being Poor* (Stockholm: Nobel Foundation, 1979); George E. Schuh, "Improving the Development Effectiveness of Food Aid," draft (Washington, D.C.: USAID, 1979).

29. Appendix 18 shows the countries that in 1978 and in the twenty-two-year period 1956–78 obtained these medium-term credits. Three countries, Poland, Portugal, and South Korea, accounted for more than half of the loans granted in 1978. With the Philippines and Peru, these five countries have received two-thirds of the loans granted from 1956 to 1978.

30. *The Agricultural Attaché, His History and His Work* (Washington, D.C.: USDA-FAS, FASM-91, rev., September 1972).

31. See Jacques Guibé, French agricultural attaché, "La promotion des exportations agricoles et alimentaires par les Etats-Unis" (Washington, D.C., October 1977); C. D. Caldwell, Canadian agricultural attaché, "Programme des associations américaines de coopérants pour le développement des marchés," (Ottawa, *Agriculture étrangère*, 1979).

CONCLUSION

1. See also the special issue of *Cahiers français de la documentation française* for May–June 1981: *Géostratégie et économies mondiales*.

2. This would mean a six- to seven-fold increase in the price of wheat.

3. See Henri de Farcy and René Groussard, "Où va l'agriculture française?" *Académie des sciences morales et politiques*, Paris, May 19, 1980, and "L'arme alimentaire des Etats-Unis," *Projet* 109, Paris, November 1976.

4. Robert Bergland, "Address to the U.S. Chamber of Commerce Conference on Results of the Tokyo Round," Washington, D.C., January 25, 1979.

5. Farcy, "L'arme alimentaire des Etats-Unis."

6. See Christian Schmidt, "Un abus de langage," *Le Monde*, October 21, 1980.

Index

About the Authors

Alain Revel is executive vice-president of GERSAR, an international agricultural and hydraulic engineering corporation in Paris. He was agricultural attaché to the United States and Canada for the French government during the 1970s. Christophe Riboud is a professor at the Agro-Food Institute in Paris and is president of IFOP, a French polling firm.